Among the Creationists

Among the Creationists

DISPATCHES FROM THE ANTI-EVOLUTIONIST FRONT LINE

Jason Rosenhouse

OXFORD
UNIVERSITY PRESS

Oxford University Press, Inc., publishes works that further
Oxford University's objective of excellence
in research, scholarship, and education.

Oxford New York
Auckland Cape Town Dar es Salaam Hong Kong Karachi
Kual Lumpur Madrid Melbourne Mexico City Nairobi
New Delhi Shanghai Taipei Toronto

With offices in
Argentina Austria Brazil Chile Czech Republic France Greece
Guatemala Hungary Italy Japan Poland Portugal Singapore
South Korea Switzerland Thailand Turkey Ukraine Vietnam

Published by Oxford University Press, Inc.
198 Madison Avenue, New York, NY 10016

www.oup.com

Library of Congress Cataloging-in-Publication Data
Rosenhouse, Jason.
Among the creationists : dispatches from the anti-evolution frontline / Jason Rosenhouse.
p. cm.
Includes bibliographical references.
ISBN 978-0-19-974463-3 (hardcover)
1. Creationism. I. Title.
BS651.R757 2012
231.7′652—dc23 2011026995

1 3 5 7 9 8 6 4 2

Printed in the United States of America
on acid-free paper

To my parents, who taught me the value of
being open-minded but skeptical.

CONTENTS

PART IV Visiting the Creation Museum, Petersburg, Kentucky, August 2010

PART V The Sixth International Conference on Creationism, Pittsburgh, Pennsylvania, August 2008

LIST OF FIGURES

PREFACE

Just over a decade ago, because of events described in part I, I developed a strong interest in evolution and creationism. Though this interest was somewhat tangential to my professional work in mathematics, I gradually found it occupying an increasing amount of my time and my thoughts. I indulged my interest in part by becoming a regular attendee at creationist conferences. After several dozen such events over a number of years, I found myself dissatisfied with much of the scholarly literature in this area. I felt there was a story to tell in what I had experienced and that I could paint a realistic picture of why creationists believe the things they do.

This book has three main purposes. It is a memoir recounting some interesting experiences I've had socializing with people whose worldview differs greatly from my own. It is also an explication of the beliefs and attitudes that are common in the anti-evolution subculture. And it is a discussion of certain questions about the relationship between science and religion that arose naturally through my experiences. The book is structured as follows:

Part I contains some anecdotes intended to whet your appetite. In the first three chapters I present a handful of stories that are broadly typical of my experiences among the creationists. As described in chapters 1 and 2, I found that insularity and scientific ignorance were distressingly common. That aside, as described in chapter 3, in most cases I was treated respectfully, and sometimes even generously, by my fellow conference attendees, even after it was clear that I did not share their views. Chapters 4 and 5 present some relevant biographical information about how I developed my interest in this subject and about my general views toward science and religion.

Part II reports on the Creation Mega-Conference. This was a large, week-long conference held in Lynchburg, Virginia, in 2005. It shall serve as our introduction to young-Earth creationism, which is what most people think of when they hear the word "creationism." After providing a primer on evolution in chapter 6, chapters 7 and 8 consider basic aspects of creationism and clear up some misconceptions about creationists that have been common in my experience. Chapter 9 discusses creationist jargon. Chapters 10, 11, and 12 consider, and find wanting, certain creationist scientific claims, drawn from biology, geology, and mathematics. Part II closes, in chapter 13, with a consideration of how creationists view popular culture, especially as presented in movies and television.

Part III focuses on a one-day conference called "Darwin vs. Design," held in Knoxville, Tennessee, in 2007. The proceedings were devoted to intelligent design

(ID) theory, which became a major player in anti-evolutionism during the 1990s. Intelligent design sheds much of the baggage of traditional creationism, taking no stand on the age of the earth, Noah's flood, or questions of proper biblical exegesis. Instead it argues simply that fully naturalistic processes are insufficient to explain the complexity of organisms. Central to understanding ID is appreciating that religious objections to evolution go far beyond concerns about the proper understanding of the Bible. Chapter 14 illustrates this by considering some of the religious responses to evolution in the late nineteenth and early twentieth centuries. Intelligent design must also be understood in the context of the legal history of disputes about evolution in U.S. public education, discussed in chapter 15. Two further chapters look at certain rhetorical techniques that are common in ID literature: Chapter 16 focuses on the unsavory practice of quoting people out of context. Chapter 17 discusses the role of conversion stories in ID discourse. The frequency with which my fellow conference attendees have told me about their own conversion and religious experiences prompted me to consider that topic in chapter 18. Chapters 19 and 20 turn to philosophical questions about the nature of science. Chapter 21 provides a brief discussion of the most significant ID biological argument.

Part IV describes what I found during several visits to the Creation Museum, which opened in 2007 in Petersburg, Kentucky. This is a professional, commercially successful venture that presents the young-Earth view of biblical exegesis and science. The centerpiece of the museum is a labyrinth leading you through Genesis 1–11. Chapters 22, 23, and 25 provide a whirlwind tour through the museum. In chapter 24 I briefly pause the tour and, inspired by exhibits about Adam and Eve's disobedience, consider the problem of evil. After completing the tour it is almost impossible to avoid thinking about how to interpret Genesis. I argue, in chapter 26, that for all the genuine problems evolution presents for Christianity, conflicts with the Bible should not be considered among them. We close this section with two chapters of a more personal nature. In chapter 27 I discuss why I am unimpressed by attempts by certain religious scholars to preserve the historicity of Adam and Eve, and the closely related doctrine of original sin, in the face of the challenge posed by modern science. Then, having devoted so many pages to the factual assertions that are sometimes made on behalf of religion, I devote chapter 28 to a celebration of cultural religion.

Part V opens in chapter 29 with an introduction to the Sixth International Conference on Creationism, held in Pittsburgh, Pennsylvania, in August 2008. This represents the most serious attempt by creationists to hold an actual scientific research conference. Chapter 30 contains a discussion of one of the greatest areas of tension between evolution and Christianity: the question of human significance. In particular, I focus on the question of whether humanlike creatures can be seen as an inevitable end result of evolution, or whether our appearance was an historical accident unlikely to occur again on a second "play of the tape." The final chapters close the book in the same way we started, by focusing on stories and anecdotes.

My conclusions will not be to everyone's liking. In particular, I do not agree that evolution and Christianity can easily coexist. My claim is not that there is a flat-out contradiction between the two, but simply that evolution provides enough to disquiet a traditional Christian that it is unsurprising that so many believe they must choose one or the other. Moreover, it is a mistake in my view to think that anti-evolutionism is primarily about an idiosyncratic interpretation of the Bible clung to by a handful of extremists. Of far greater concern are the implications of evolution for human significance, the cruelty of the evolutionary process, and the demise of the traditional design argument. In my conversations with creationists, these are the points, far more than concerns about the age of the Earth or an understanding of Genesis 1, that seem endlessly to recur.

Many people defend theistic evolution, by which I mean a system of thought that accepts both traditional conceptions of God and modern evolutionary science. The literature in support of this view is vast, and we shall only be able to discuss a small portion of it here. I hope, though, that I have given adequate consideration to some of the major lines of argument and have explained why so many anti-evolutionists, not to mention so many nonbelievers, find them unpersuasive.

I have tried to assume as little background as possible, but it has not been practical to assume nothing at all. Thus, I will take for granted a basic familiarity with the stories in the first eleven chapters of Genesis. At times I will refer casually to the days in Genesis 1, to Adam and Eve, to Noah's flood, or to the Tower of Babel, and I will assume that you know what I am talking about. Except where otherwise indicated, all Bible quotations come from the New Revised Standard Version. Also, though I give a brief introduction to the major claims of evolutionary theory in chapter 6, a reader with absolutely no prior knowledge at all of this subject might find certain sections to be tough going. These sections are few, however, and the most glancing familiarity ought to be sufficient for most of the book.

In certain chapters I present quotations from the speakers at various conferences. You can assume that any quotations presented without an explicit source being given have been transcribed from audio or video recordings of the event. In some cases the speakers were delivering presentations that were also available on DVDs. In some of these cases, indicated in the endnotes, I relied on the DVD version for direct quotations. All of the photographs were taken by me.

I should also emphasize a few things this book is not: First, it is not about presenting the evidence for evolution. I write from the perspective of someone who finds evolution persuasive, but I am not necessarily trying to convince you of that fact in this book. Second, this is not meant to be a comprehensive treatment of creationist claims. I will address a few of their major arguments, but this is not primarily a book of the "Creationists say *X*, but the reality is *Y*" variety.

The evolution/creation issue covers a lot of terrain, some of which is left completely unexplored in my discussions. I make little mention of the politics of school board disputes, for example. Nor do I attempt to formulate clear strategies for moving people away from creationism, or from anti-science views more generally. Proper treatments of those topics would simply require far more space than I can grant them here. For the purposes of this book, I am more interested in understanding the problem than I am in formulating solutions.

Finally, this is not a polemic. By the end of the book, no doubt will be left as to where I stand on a variety of contentious issues, but none of my criticisms should be taken as personally acrimonious towards anyone, including the creationists. I have my opinions, but I do not take it amiss when others disagree with me.

This book has benefited greatly from the heroic proofreading efforts of several people. Glenn Branch, of the National Center for Science Education, is close to omniscient on the evolution/creationism issue. He directed me toward many interesting references, helped me avoid several embarrassing factual errors, and showed me numerous ways of sharpening my arguments. I am also deeply indebted to Jerry Coyne of the University of Chicago, who commented on every portion of the first draft and gave me many helpful suggestions. Also providing helpful feedback were Bernadette Barton, Ophelia Benson, Linda Coleman, Douglas Gill, David Neel, Eugenie Scott, and Laura Taalman. This book was greatly improved by their suggestions. Of course, the fault for any remaining errors lies entirely with me. It is also my pleasure to thank Phyllis Cohen, of Oxford University Press, for her extraordinary patience and encouragement while I was working on this project.

PART I

Vignettes

1

Studies in Insularity

In August 1999, the State Board of Education in Kansas voted to implement science standards that excluded evolution and the Big Bang. The result was a barrage of hostile press coverage and a competition among pundits for the choicest anti-Kansas put-downs. Stephen Jay Gould spoke for many when he wrote:

> Why get excited over this latest episode in the long, sad history of American anti-intellectualism? Let me suggest that, as patriotic Americans, we should cringe in embarrassment that, at the dawn of a new, technological millennium, a jurisdiction in our heartland has opted to suppress one of the greatest triumphs of human discovery. (Gould 1999)

At that time I was a fourth-year graduate student studying mathematics at Dartmouth College. Though I was vaguely aware of the goings-on in Kansas, I never imagined they would have any effect on my life.

Then, in March 2000, I accepted a postdoctoral position at Kansas State University (KSU). The Board of Education had also implemented new mathematics standards the previous year, and part of my new job involved integrating those standards into KSU's teacher certification program. In doing this work, I met many people involved with public education in Kansas, including some on the frontline of the evolution dispute.

Shortly after my move I learned of a conference for religious home-schoolers to be held in Wichita, the largest city in Kansas. The featured speakers were from an anti-evolution organization called Answers in Genesis, though that name meant little to me at the time. Since anything related to education in Kansas had relevance to my job, and since Wichita was a reasonable drive from my home, I decided, on a whim, to attend.

While at the conference I attended a session addressing the practical problems of starting a home-school. An issue that arose repeatedly, both from the presenters and from the audience during the question period, was that of protecting their children from the perverse influence of "the culture." It was clear that, from their perspective, evangelical Christianity was a tiny island of righteousness adrift in a sea of secular evil. For these parents, sending their children to a public school was tantamount to placing them in a war zone. They were trying to steer their children's souls to God, while Satan, using tools like Bible-denying science, acceptance of homosexuals, and secular humanist philosophy, had other plans.

Near the end of the question period, I asked the speaker if he ever worried they are being *too* protective of their children. Eventually the kids will leave home and go out into the world. They are inevitably going to encounter all those ideas the parents find so threatening. Perhaps it is better they encounter them under circumstances where their parents still have influence.

His reply was wonderfully blunt. He said, "No, we don't worry about that."

The keynote presentations were a strange mix of science and revival. A speaker would present an obscure and technical point in geology, and various audience members would yell out "Amen!" or "Praise God!" Prayers were ubiquitous, as were exhortations to beware of secular, Bible-challenging science. We were told that "true science" confirmed creation, not evolution. To judge from all the nodding I would guess the message was well received.

One of the speakers was Ken Ham, the president of Answers in Genesis. He found it annoying that Christians more moderate than he were willing to "compromise" the plain teaching of the Bible. "Sometimes 'day' means DAY!" he thundered, rebuking those who argue that the days in Genesis 1 are just figures of speech or refer to long periods of time. He also made forays into science, at one point making free use of the mathematical theories of probability and information. The audience was enthusiastic, but I was less impressed. It seemed clear to me that his arguments ran afoul of certain basic facts.

Some of the talks had formal question-and-answer periods, but Ham's did not. Instead he remained at the front of the room so that folks could talk to him. I joined the crowd of roughly fifteen people, and when it was my turn to speak I pointed out some elementary flaws in his arguments. I did not raise my voice or hurl any insults, but I also did not allow his answers to go unchallenged. Some of the others in the crowd asked me questions, with varying levels of politeness, and the conversation went through several rounds. Clearly impatient, Ham eventually recommended a book on sale at the conference bookstore that he said would answer my questions. I thanked him and left.

I bought the book and read enough of it over lunch to realize that its arguments were as fallacious as Ham's. Later that afternoon, I noticed Ham walking by himself across the convention floor, and I fell into step beside him. I showed him the book and explained that it did not respond to my objections. This time, alas, the conversation got heated.

As we argued it became painfully clear that he simply did not understand the topics he was discussing. Eventually he told me I was very arrogant and needed to learn more about the subject. I replied that arrogance was standing on a stage pretending to know something about science.

For some reason that ended the conversation.

* * *

In 2003 I left my job at KSU to accept a position at James Madison University in Harrisonburg, Virginia. My new home also provided ample opportunities to attend creationist gatherings, and in June 2010 I was present at another conference for religious home-schoolers, this time in Richmond. While at the conference, I attended a session called "What You Need To Know About College, But Probably Don't." We were warned that college professors are committed to indoctrinating helpless students with left-wing propaganda. Ominous statistics were presented documenting the scourge of professorial liberalism, with special emphasis on the widespread and, in the opinion of the speaker, horrifying acceptance of homosexuality. As accustomed as I now am to this sort of thing, I remain astonished at the gall of conservative Christians lecturing others about indoctrination.

The natural milieus for creationism and its supporting ideology are the stage and the pulpit. Given a platform for unfettered dudgeon its leading proponents always rise to the occasion. So long as the audience is generally sympathetic and there is no opportunity for counterpoint, its confident fanaticism is infectious. No matter how cool and rational you think you are, no matter how well educated in science and history, I defy you to listen to its most eloquent champions without becoming at least slightly entranced.

Placed instead in a forum of calm deliberation and rigorous evidential standards, creationism becomes far less impressive. Shorn of its emotionalism, nothing remains beyond its gross scientific errors. That is why creationism's most notable failures have come in the courtroom and the seminar hall. It is also why creationists invest so much energy in demonizing judges and professors.

The picture of dogmatic science professors relentlessly indoctrinating their free-thinking students is a near perfect inversion of reality. Most professors are delighted when their students ask challenging questions, so long as they are not obnoxious or rude. It shows they are engaging seriously with the material, which is what we really care about. From the other side, most students, regrettably, are all too happy to be indoctrinated. They will recite whatever catechism we hand them if they think it will get them a decent grade.

Education does pose a threat to creationist religious faith, but not because of any supposed dogmatism on the part of professors. The true explanation is far simpler. Exposure to novel and varied ideas, and to classmates from a variety of religious backgrounds, inevitably dissolves the cocoon in which creationists too often enclose themselves. It is in the nature of simplistic, totalizing philosophies that they wither before the messiness and complexity of the real world.

At the Richmond conference, mathematicians came in for some criticism in a talk entitled "Math from a Biblical Worldview?" The speaker made it her task to refute the false and pernicious notion that our understanding of mathematics is independent of our views about God. She used a number of quotations to illustrate the basic problem, such as this one from physicist Heinrich Hertz: "One cannot escape the feeling that these mathematical formulas have an independent existence and an intelligence of their own, that they are wiser than we are, wiser even than their discoverers, that we get more out of them than was originally put into them."

"Where is he giving the glory?" she asked. After being greeted by silence, she prodded us by emphasizing the phrase, "that they are wiser than we are." A few people caught on at this point, but I was not one of them. Then came the dramatic answer. "To the math itself," she said sadly, to grumbles of approval from the audience.

I would have thought Hertz's statement could more plausibly be seen as supportive of God than as a repudiation. One suspects that, by suggesting that the formulas were wiser than their discoverers, Hertz, a practicing Lutheran, was honoring the Mind he saw behind the math. The speaker's style of argument here is a commonplace of fundamentalist discourse. Omitting direct reference to God is treated as equivalent to rejecting Him.

After the talk I purchased a copy of her book. As a math teacher myself, I am always interested in novel approaches, but much of what I found struck me as very odd. For example, in a chapter entitled "Harm to the Heart," she writes:

> I remember the confusion I felt as I entered high school. Math seemed like a big mystery to me. Why did math work? Where did math come from? My textbooks never really told me. I found myself memorizing rule after rule without really understanding how that rule came about. I began losing sight of the purpose behind learning math. How would I ever use exponents and algebraic division/graphing in my own life?
>
> Had I understood that math was not neutral—that math merely records the order God created—math would not have been a mystery to me. I would have realized that the rules in math are merely ways of writing the complex principles by which God holds everything in the physical universe in place. Had my textbooks presented math biblically, they would have taught me how exponents and algebraic division/graphing are useful, God-given tools. They would have shown me that math has a meaning and a purpose.[1] (Loop 2005, 36)

Not understanding the reasons for some mathematical rule is a frustration known to all students of the subject, and I agree that textbooks routinely do a poor job of explaining matters. That said, it is hard to imagine what mystery encountered in a math class is resolved by invoking God's handiwork. If you are theistically inclined, you might view the general orderliness of the universe as a reflection of divine grace. Perhaps you find in this a satisfactory explanation of why there are effective mathematical rules at all. But this perspective certainly will not tell you the reasoning behind the specific rules we use, or how those rules are applied in practical situations. I have been teaching college-level mathematics for over a decade, and during that time I have been asked thousands of questions by confused and frustrated students. I cannot think of a single one to whom it would have been helpful to reply, "Because God willed it to be so."

This is merely one item. On virtually every page of the book, I found evidence of an overly simplistic understanding of mathematics. I barely recognized my discipline in the presentation. This is a common experience for scientists who immerse themselves in the creationist subculture. Creationists speak with confidence on a great many branches of science, but do so in a manner that seems confused and misinformed to professionals.

But even more than evidence of scientific error, what struck me at the Richmond conference—as at other evangelical and creationist conferences I have attended—was the incredible insularity of my fellow attendees. As they saw it, if their children went their whole lives without ever hearing about evolution or about views of morality different from their own, that would be no loss whatsoever. Above all else their faith must not waver. To enter the convention center or hotel ballroom of these conferences is to experience a profound sense of vertigo, as though the rules really have changed just by passing through the door. With everyone agreeing with everyone else and speaking with such confidence, it is very difficult to keep your wits about you.[2]

2

Showdown at Subway

In March 2007, I was in Knoxville, Tennessee, for a conference called "Darwin vs. Design." It was devoted to intelligent design, a softer version of creationism born of various pro-evolution court decisions during the 1980s. The conference site was the Knoxville Convention Center, which, regrettably, offered few affordable lunch options within walking distance. Those of us who had not thought to bring a sandwich mostly availed ourselves of the only convenient place to eat: a nearby Subway sub shop. More than one hundred people packed a small eatery staffed by two teenagers plainly not expecting so large a lunch crowd.

I was on that line, contemplating the very long wait between me and my turkey on wheat, when I happened to overhear a conversation among the people immediately in front of me. There were four of them, an older woman and three teenagers. Initially I thought the woman, call her Anne, was the mother of the three teens, but later I learned they were not related.

Anne was quite animated. "Did you see those people with the 'Happy Atheist' shirts?" she asked, with scarcely concealed disgust. "They were handing out leaflets."

I perked up. I had not noticed those people. If I had, I would have lent them some moral support.

Anne continued: "I wish one of them were here. I can't imagine how they respond to all the evidence for design."

I took that as an invitation and said: "Well, I'm an atheist. Why don't you ask me?"

She was taken aback, but the teens were immediately interested. I soon learned that the oldest was nineteen and a student at the University of Tennessee. The other two were high school students, one boy (sixteen) and one girl (fourteen). The sixteen-year-old ended up monopolizing much of the conversation.

Things started off cordially when the college student asked me how I responded to the arguments made in the morning talks. I replied that for the most part the speakers were not providing the full story. In many cases their factual assertions were simply false, and their arguments, for all their superficial sophistication, did not hold water. I focused especially on certain mathematical arguments presented in the talks.

The sixteen-year-old challenged me on these points, and the ensuing conversation was both interesting and frustrating. Interesting because he was clearly very bright and engaged, but frustrating because he was stubbornly confused about certain mathematical issues. This raises a difficulty I have experienced on many occasions.

Imagine you are in the audience during a scientific presentation, and the speaker flashes an equation on the screen. He tells you that careful study of the equation reveals something significant about the natural world. If you have some training in science or mathematics, there are certain questions you will ask instinctively: Given that any such equation is based on an abstract model of reality, you will want to know whether the assumptions underlying the model are sound. You will wonder whether the variables, which are said to represent real-world quantities, can be measured, or at least approximated, with reasonable accuracy. And, from a purely mathematical standpoint, you will want to know if the algebraic manipulations are correct. The arguments made by proponents of intelligent design routinely fail all three tests, which is why mathematicians and scientists typically respond to them with disgust.

Most people who attend creationist conferences have no such training. They have instead a deep intuition that evolutionary theory is suggesting something highly improbable, if not flatly impossible. Our everyday experience tells us that things break down and decay when left to the mercies of nature—yet here comes evolution to claim that natural forces have not only caused organisms to become more complex over time, but have actually crafted functional structures like the human eye and the wing of a bird. It seems awfully far-fetched. The result is that they do not, upon seeing the equation flashed on the screen, think carefully about the perils of abstraction, the practical difficulties of measurement, or the details of algebraic manipulation. Instead they see only that their commonsense intuition seems to have scientific backing.

Though we talked about science for a minute or two, the conversation quickly turned to religion, morality, and the Bible. At one point Anne asked me bluntly if I believed in history.

Smelling a trap, I cautiously replied that I did, indeed, believe in history. "Well," she said, "then you should believe in the resurrection of Jesus. That's history!"

"Is that so?" I asked.

"More than five hundred people witnessed it!" she replied. "How can you deny it!"

This is another common feature of creationist argumentation: that the Bible asserts something to be the case is reason enough to assume it is true. In this case the reference is to Paul's statement in 1 Corinthians 15:6, "Then he appeared to more than five hundred brothers and sisters at one time, most of whom are still alive, though some have died," where "he" refers to the resurrected Jesus. I replied that five hundred independent accounts of an event is strong evidence that it occurred, but one person telling me what five hundred others saw is far less impressive.

The conversation ranged over a great many topics.

How could consciousness be explained without reference to an immaterial soul? Why are there no transitional forms in the fossil record? Where did the universe come from? How could DNA just have created itself? What about all the evidence of cosmological fine-tuning? You think it's all an accident? I fielded the questions as best I could. All of these are standard talking points, but being outnumbered four to one I was rarely able to utter two consecutive sentences without being cut off.

Somewhere around here a new player entered the game. I was in the middle of some staggeringly eloquent and totally persuasive point about genetics or whatever when the woman immediately behind me on line, I will call her Jane, decided she had heard enough.

She cut me off and spoke directly to the three teens. Without making eye contact or acknowledging me in any way, she said something like, "I just want to say that I think you three are beautiful children and that it's very important that you read the King James Bible and that you beware of people who have been educated beyond their intelligence." That they were "beautiful children" and should be reading the King James Bible were points she made repeatedly.

As for the part about people being educated beyond their intelligence, it's possible that was directed at me.

I have been involved in conversations of this sort at most of the conferences I have attended. As long as there is only one evolutionist among a large number of creationists the majority does not feel threatened and instead treats the errant Darwinist like some sort of zoo animal. I am still uncertain as to the best way of handling this situation. My instinct, upon being peppered with long-refuted canards from people wrongly convinced of their own erudition, is to unload with both barrels. Generally, though, I think this is counterproductive. I have tried that approach on several occasions, but I have always regretted it later. However temporarily satisfying it is to fire a real zinger, any hope of doing long-term good comes from being scrupulously polite. I have already remarked on the insularity accompanying the strongest forms of fundamentalist beliefs. It is very possible I was the first actual evolutionist Anne and Jane had met in quite some time. So let them see someone who on the one hand is completely unafraid of any challenge they might offer, but who also has no desire to be insulting or aggressive.

In a way I was flattered by Jane's interruption. Her desperate tone suggested she feared I was having some impact on the beautiful children. When she finished, I feigned offense and asked, "Are you suggesting that I am not beautiful?" She laughed and told me that I, too, was beautiful and that she prayed that some day I would come to know the "Truth" (a term that I am pretty sure she intended to be capitalized).

Happy for the support, Anne moved closer to Jane. They discussed me in the third person. I found that rude, but took the opportunity to return my attention to the teenagers. We resumed where we had left off.

At one point Anne, having absorbed some of Jane's energy, asked me pointedly why I found it so hard to believe in God. I asked her why she found it so hard to believe in unicorns. The college student thought that was funny. Anne did not.

I then suggested that you do not really explain anything by invoking God. Nonsense, she replied. God explains all sorts of things. I said I did not agree. She was just taking everything that was mysterious or hard to understand in the world and placing the label "God" on it. That's not an explanation.

I asked her, "If you think God explains how the universe came into being, then tell me how He did it."

"He spoke the world into existence."

"Just like that?"

"Just like that."

"He said let there be light, and there was light?"

"Yes."

"And you find that *easy* to believe?"

"Yes."

Then things got *really* weird. Jane grabbed my hand and claimed my soul for Jesus. I was too shocked to point out that my soul was not hers to claim. Instead I thanked her awkwardly and tried to get back to the teenagers. She seemed disinclined to release me, however, and went on to explain that she knows that God will grant me an awakening and that I will come to Jesus and that we will meet again in heaven and that I will know her name when I see her there. I offered that anything is possible and redoubled my efforts to extricate my hand.

At some point, a genuine miracle occurred! It was my turn to order.

We are now approaching the story's dramatic conclusion, but to get the full effect I will need to reveal a personal detail. I very much dislike cheese. I never touch the stuff, and I even find it difficult to watch other people eating it.

I ordered my turkey on wheat, but they were out of wheat. I asked for it on their "Italian" bread, but they were out of that as well. What sort of bread did they have? He rattled off a few options, all of them involving words like "Parmesan" and "Asiago."

I turned to Jane, mentioned my inability to eat cheese, and asked her if God was punishing me. She laughed and said this was just bad luck.

3

The Prospects for Agreeable Disagreements

The sixth International Conference on Creationism (ICC) was held in Pittsburgh, Pennsylvania, in August 2008. The conference sessions had an interesting format. Most were ninety minutes long with plenty of time set aside for questions. These Q&A's were as lively and contentious as anything I have seen at more mainstream conferences, and at times they were downright nasty. Mind you, they were not contentious because of people like me asking skeptical questions. I am talking instead about disputes among the creationists themselves. These lively interactions were generally encouraged.

I was in the audience for a talk bearing the formidable title "A Nuanced Lakatos Philosophy of Theology and Science." The speaker was Doug Kennard, then a professor of biblical studies at Bryan College in Tennessee. His presentation was hardcore philosophy of science, very little of which had anything specifically to do with creationism. The talk was dense and technical and mostly would not have been out of place at a university philosophy colloquium.

Kennard encouraged creationists to adopt a more nuanced view of the philosophy of science, arguing they would thereby be more likely to gain respect for their views. In the paper accompanying his talk, he wrote, "Thus a creationist model needs to focus on crafting better theories rather than merely trying to falsify evolution or they will never be seriously considered by the academy as a live option. Attempts to falsify evolution will be simply considered as annoyances like mosquitoes from quacks like flat earthers" (Kennard 2008, 34).

There is some truth in this. Long-standing, highly successful, scientific theories are not abandoned because someone points out a small anomaly in the data. The trouble is that Kennard strongly implied that evolution had been falsified by creationist arguments. This is simply untrue. Creationists are not even successful in pointing to difficult questions about evolution. During the ensuing question period, I went to the microphone and said, verbatim, the following:

> I'd like to address one narrow point you brought up—and it came up again moments ago in your answer to a previous question—this issue of how creationists can present themselves in ways more likely to be received by the

scientific community. I think, to be a little blunt about it, I don't think the issue is philosophy. I don't think it's that scientists are operating within a Kuhnian paradigm or something like that. I think it's simply that from the scientific perspective creationists are just making bad arguments. That's how they see it. When they look, and I should stop saying they because I include myself in this, when I look at a lot of the books out there and browse through them, you can go pages at a time without seeing anything that resembles the way scientists talk about evolution. And I would also add that a lot of these arguments *have* been addressed by scientists. It's not that they've been ignored and have not been able to get a hearing. Especially more recently with people like [ID proponents] William Dembski and Michael Behe, you have book-length arguments about it. So I think that philosophy is not really the answer. I think it's that there are a lot of bad arguments.

And worse than that, when I hear, when I attend conferences like this and I hear creationists say, "Oh, they're so disrespectful and they're so rude," well, I haven't seen much respect coming the other way. When I attend these conferences, I hear caricatures of science, I see misquotations of scientists' work, and very simplistic versions of evolution. I would suggest, instead of finding comfort in philosophy, sharpen your arguments and be a little more respectful the other way. Then they might be more inclined to reciprocate.

I will come to Kennard's reply in a moment, but first there is a different point to make. You might think that in an audience of creationists my little piece of oratory would have been met with hisses and sneers. You would be wrong. No one expressed any disapprobation at all. No one tried to shut me down or to get me away from the microphone.

People often ask if I have ever felt physically threatened at a creationist conference. The answer is no, never, not even once. It is rare (though it *has* happened) that I am treated rudely in any way at all. A far more common reaction is to have people approach me later to ask me to clarify my views, or to ask me questions. (Invitations to Bible studies are also quite common.)

Kennard's reply went on for several minutes. Here's how it began:

Well, I agree and think those are issues that need to be worked on. Having gone to the Mendel computer-model talk, I wonder if that's not a good example of what I'm urging here in the Lakatos approach. . . . That's one side of this approach that I think might say that, it's not just that we haven't answered, or that we don't have good scientific presentations. I think that a lot of the literature out here is purely at a popular level, and probably some of it is not good science.

I had not expected such forthcoming remarks, but that was not the only surprise.

When Kennard finished, the session moderator went to the microphone and said, "Thank you for coming. We need the challenge and we need the sharpening, the honing."

The moderator was a retired air force pilot, and he caught up with me after the talk. He thanked me again, and told me that the point of the ICC was to provide a forum for serious work in creation science, as distinct from the popular-level stuff that gives creationism a bad name. He told me about all of the garbage papers they received for inclusion in the conference and how difficult it was to single out the few good papers from the piles of nonsense. He specifically used the phrases "garbage papers" and "piles of nonsense." I was a bit surprised by his honesty and bluntness. We chatted for several minutes, and it was all very pleasant and fruitful. It is an experience I have had many times, which must be one of the reasons I keep attending these conferences. It is hard to imagine any view of the world less appealing to me than Christian fundamentalism, and fundamentalists look at my atheism with comparable disdain. Sometimes, though, we have moments like this. Moments when it seems a real possibility that, while a meeting of the minds is surely asking too much, we can at least disagree in an agreeable manner.

That was just one of many pleasant experiences I have had at these conferences. An especially memorable one occurred at an intelligent-design conference held near Kansas City in 2003.

A group of seven folks was driving to a local Lebanese restaurant for dinner, and they invited me to join them. When I asked if I first needed to make sure I had a ride back to the conference, they laughed and assured me everything was on the up and up.

Since I was not very hungry and was ordering only to be polite, I scanned the menu for the lightest item I could find. That turned out to be a vegetable kebab, which I duly ordered. In retrospect, that might not have been the strategic choice. They already thought I was a liberal, godless lunatic. Now they would think I was a vegetarian as well.

Eventually the pressure of a seven-on-one conversation wore me down, and I started looking for ways to redirect the flow. There was a high school student among us, and at one point he asked how I could explain the survival of the Jewish people over the centuries if God were not looking out for them. As a Jewish person myself, my first thought involved the Holocaust, but going that route seemed like a poor way of lightening the mood.

So without looking up I deadpanned, "That's easy to explain." I removed a cherry tomato from my kebab skewer.

Puzzled, the student asked, "How do you explain it?"

I looked him dead in the eye and said, "It's because the Jews controlled the banks."

I am happy to report that the student had no idea what I was talking about. The other folks, however, laughed uproariously. One of them called out, "And the entertainment industry!" and more laughter ensued. The conversation quickly moved on to topics unrelated to science, religion, or politics, and I think we were all happier for the change of pace.

* * *

People often tell me I am wasting my time at these conferences. "You're not going to convince anyone!" they say, as though informing me of a possibility I had not considered. I have no illusions, or ambitions for that matter, that anyone is going to slap his forehead in response to some super-clever argument from me.

As I suggested in chapter 2, I do hope I have managed to puncture, if only a little bit, the insularity so typical of creationist culture. Maybe I have planted a seed or two in the minds of the people I have met that science is not really as simple as their leaders are telling them. Not everything has to be the stuff of revolutions, after all. If I have managed to show just a few people that we liberal, atheist, scientist types are not really so bad, then it has been worthwhile on those grounds alone.

I do not know if I have had even that much effect, but I do know these experiences have had a salutary effect on me. It is far more difficult to caricature and stereotype people you have actually met. Have a few conversations over lunch or during breaks at conferences, and suddenly they are no longer abstractions or types. They are no longer defined by a few odd beliefs you have heard that they hold. They become actual people, with depth and personality and reasons for the things they believe.

Insularity is a two-way street. It is not for nothing that academics have a reputation for living in ivory towers. When I tell my professional colleagues about my little hobby, I know that I am likely to hear certain questions in reply: Are creationists just stupid? How can they accept science when it gives them technology and medical advances, but reject it on the age of the Earth or the origin of species? How can they put so much faith in the Bible?

Well, yes, come to think of it, why *do* creationists accept science on some questions and reject it on others? How *do* they come by their confidence in the inerrancy of the Bible? Were those serious questions or just insults in the form of questions? For me these are serious questions, and I can think of no better way of getting answers than to go where creationists are and to ask them directly.

Relax, I'm not going soft. Creationism rests on a foundation of error, no question about it. Amicable conversations are well and good, but there really does come a time when you must put your foot down and insist that facts are facts. When this all enters the political arena, especially in the domain of school board disputes, then the angry confrontations and courtroom dramas, while regrettable, are unavoidable.

But for all of that, we still have to live together. Given this simple reality, it cannot be the worst idea in the world to try talking to each other once in a while.

4

How I Got Here

If it seems strange to you that a nonreligious mathematician would devote so much time to attending these gatherings, then I assure you it seems strange to me as well. It certainly is nothing I ever imagined myself doing when I was in high school or college. My interest in this issue came gradually.

From 1995 to 2000 I was a graduate student in mathematics at Dartmouth College in New Hampshire. While I was there, the famous Christian evangelist Luis Palau came to preach. Being vaguely familiar with Palau, I was curious to see his presentation. Since I did not want to go alone, I tried to persuade some of my more religious friends to go with me. To my surprise, they seemed uninterested. Somewhat unusually in the context of academic science, most of my fellow graduate students were active Christians. I, on the other hand, was the vocal atheist in the group. Try to imagine the incongruity of an outspoken nonbeliever pressuring his believing but reluctant friends to see a world-renowned preacher. In the end, however, I was sufficiently persuasive, and two of my friends agreed to keep me company.

Palau's presentation hewed very closely to every stereotype you have heard of the Bible-thumping preacher. It was long on admonitions to find Jesus, but short on reasoned arguments. I was told to pursue a personal relationship with God, but details on how one does that were not forthcoming. I was told that such problems as I faced were the result of not putting Jesus first in my life. That struck me as implausible. Ominous references to the fate of unbelievers in the afterlife were common. This fate was presented with a level of enthusiasm that bordered on tasteless.

Palau's sermon never brought me close to a conversion, but it certainly held my interest. Most of the audience, however, was emotionally engaged to an astonishing degree. Many had their hands upthrust in gestures of supplication before God. When called upon to pray, they closed their eyes tightly, clasped hands with those near them, and affected facial expressions far more serious than any I had ever seen. Palau's jokes were met with uproarious laughter. His stories of past evangelistic successes were drowned out by cheers. Tales of encounters with anti-Christian forces elicited gasps and boos. Simply put, they were *into it.*

The most interesting moment came at the end. Palau told the audience that if his remarks had persuaded anyone to accept Jesus into their lives, then they should come to the front for advice regarding their next step. I briefly felt sorry for him, figuring that no one would be leaving their seats. I pictured a moment of awkward silence before Palau, embarrassed, moved on. It just seemed preposterous to me that anyone could be moved to so fateful a decision based on a preacher's ravings.

You can imagine my surprise when more than a dozen people approached the stage. I raised an eyebrow. I had listened to Palau's hour-long speech with interest, but without a trace of emotional response. The success of Palau's altar call made me wonder, not for the last time, whether there was something wrong with me for feeling nothing.

That's when one of my friends said, with the weary air of someone who had seen this many times before, "I hate this part, pressuring people to make a public display of faith." For him, religious faith had nothing to do with whipping people into emotional frenzies and taking advantage of them while in that state. It certainly was not about simplistic bromides or ostentatiously advertising your status as "one of them." Instead it was a reflective process that owed as much to the rational side of the brain as anything we were learning in our math courses. I started to understand why my friends had been reluctant to attend this little show. Palau's histrionics were, for them, a caricature of Christianity. Whereas I just found the proceedings silly, they found them offensive. It was my introduction to two very different facets of modern American Christianity.

That might have been the end of things, but for a second incident that happened soon after. The Dartmouth student newspaper published a vitriolic, anti-evolution essay in its opinion section. Evolution, according to the author, was not merely fallacious as a scientific theory, but was actually a tool of Satan. Evidence supporting it was nonexistent, but scientists, brimming with atheistic fervor, were blind to that fact. There were also some purely scientific arguments, one involving the second law of thermodynamics, another involving the complexity of the human eye.

This was all new to me. I was vaguely aware that evolution was a source of controversy in public schools, but I knew nothing about the specifics of the debate. I had friends who were studying biology, and I knew that while not all of them were atheists, all of them detested creationism. On the other hand, among my fellow mathematics students were two who were viscerally hostile towards evolution. Both were very religious, but when discussing their opinions they argued entirely on scientific grounds.

I decided to learn more about evolution and creationism. Approaching the matter in classic academic fashion, I acquired a large stack of pro-evolution books and articles, and an equally large stack of anti-evolution books and articles. Over the next several months I read every word of every one of them.

Having had the benefit of a public school education unencumbered by religious interference, I knew the basics of Darwinian evolution. I therefore began with the anti-evolution literature. I encountered a wide array of scientific arguments, drawn not just from biology but also from geology, paleontology, physics, genetics, and many other fields besides. For all I had heard about crazy religious fanaticism, I was surprised to find many of these arguments based entirely on science. There was a superficial plausibility to these assertions, and I had no ready refutations to offer.

I have never been inclined to take the Bible literally, but it seemed clear that the scientific question of evolution's legitimacy was entirely separate from the religious question of the Bible's divine origins. That is, one could reject evolution as poorly supported by evidence without necessarily accepting any particular religious view. With that distinction in mind, I was initially impressed by the creationists. I considered seriously the possibility that biologists, precisely because they were so often attacked and criticized, had overreacted by exaggerating the strength of their own case.

There were, however, two things that bothered me. The first was the relative simplicity of creationists' claims. For all their liberal use of scientific jargon, they presented little that could not be readily explained to a bright high school student. It seemed unlikely that a thriving, professional area of modern science could be undone so easily. The second was the sheer magnitude of their allegations. In their telling, evolution was not merely wrong, it was *ridiculous*. It was not simply that the evidence for evolution was weaker than was claimed, it was that there was *no evidence at all* for evolution and *massive, irrefutable evidence* for young-Earth creationism. Evolution, they would allege, was not even about science at all. It was about *promoting the religion of secular humanism*. I use the italics to capture the passion with which they made these points.

The portrayal of evolutionists throughout this literature was of benighted and confused scientists too blinded by their anti-religious prejudices to notice elementary logical fallacies in their theory. This was hard to accept. It is one thing to suggest that the scientists are wrong, but it is quite another to suggest that they are stupid or have overlooked simple errors. Still, with no responses immediately at hand, I was not inclined to dismiss creationism too casually.

That all changed the first time I encountered their treatment of mathematics.

As I have mentioned, creationists often use arguments drawn from probability and information theory, which happened to be subjects about which I knew a great deal. My knowledge of fossils and genetics was rudimentary at that time, but I certainly knew a bad probability argument when I saw one.

Just to be clear: I am not talking about subtle oversights or minor errors. I am not saying that creationists had interesting points to make, but had misunderstood some difficult, technical detail. I am talking instead about errors indicative of a total incomprehension of the subject. That got me thinking. If their treatment of subjects about which I was knowledgeable was so inadequate, what confidence could I have in their treatments of other disciplines?

Around this time I delved into the anti-creationist literature, and that was the end of any optimism I had for the scientific seriousness of creationism. A number of first-rate scholars, such as philosopher Philip Kitcher, paleontologist Niles Eldredge, and biologist Douglas Futuyma, addressed at length the arguments of creationists and the evidence for evolution, and I am grateful to them for having done so (Kitcher 1982; Eldredge 1982; Futuyma 1995). Their refutations of the anti-evolution arguments were convincing and satisfying. Moreover, I came to appreciate the extent to which the creationists had distorted both the evidence for evolution and the views of scientists regarding it.

5

My Problems with Religion

A while back my mother told me a story from my childhood. When I was around five, I asked her, out of the blue, what God was. She was a bit surprised, but replied in a serious and age-appropriate way. Apparently I responded by giving her a funny look. Then I said, "That's silly, Mom!"

During the three years I lived in Kansas, I discovered a fundamentalist radio station. One day I was listening to a call-in show for people seeking parenting advice. A caller early in the show told an awful story of her teenaged son being killed in a car accident. A later caller, referring to this, said something very close to the following: "I have also recently suffered something heartbreaking. Unlike the previous caller my son was not killed, but to me it feels just as permanent. He called home from college the other day and told me he had become an atheist."

If I am to be your host for the next two hundred pages or so, I may as well tell you where I am coming from. No one approaches a subject as emotional and contentious as science and religion without a set of biases and preconceptions, so it is only fair I tell you mine. For me, religion is simultaneously fascinating and frustrating, for reasons that shall become clear momentarily.

Let me stress at the outset that it is not my intention here to develop a comprehensive philosophical argument for the views I profess. Some of the issues I raise shall be dealt with more carefully elsewhere in the book, but this chapter's main purposes are simply to tell you what I believe and to give some indication of why I believe it. Let us begin.

I am an atheist. That means that I do not believe in God. It does not mean that I am metaphysically certain there is no God, that I wallow in nihilism and moral relativism, that I think science has explained everything, that I think religious people are stupid—or that I partake in any of the other asinine caricatures of atheistic belief you may have heard.

Addressing Christianity specifically, science plays three roles in my thinking. First, many of the specific facts that science has uncovered, evolution in particular, strike me as very difficult to reconcile with a Christian view of the world. Christianity traditionally teaches that human beings are the intentional product of a God of infinite love and that He created the world specifically so that we might live. Science tells us we are incidental by-products of a lengthy process of evolution by natural selection and that this process is wasteful, violent, and even cruel. It is logically possible these are two aspects of the same reality, but I can see no good reason for finding that plausible.

Second, following in the tradition of those nineteenth-century natural theologians who believed the existence and attributes of God could be validly inferred from a consideration of nature, I believe it is possible in principle to find scientific evidence strongly suggestive of an intelligent designer, if not the Christian God specifically. The protestations of creationists notwithstanding, science has found no such evidence. I find that significant, since I would think that the natural world is the obvious place to look for evidence of a supernatural designer. It is true that an absence of evidence is not necessarily evidence of absence, and equally true that in principle there could be lines of evidence falling outside what most people would consider science. The fact remains, however, that the failure of the natural world to provide evidence for the supernatural must surely count as a blow against the claims of traditional theism.

Third, and separate from these considerations, is the general approach to knowledge on which science is based. Science takes as its ideal that all theories must be judged by evidence. It rests on a foundation of skepticism. There is ultimately no room for proof by authority. If you take this sort of thing seriously, especially if it is the style of thinking employed in your daily work, then it becomes difficult to think of faith and revelation as reliable routes to knowledge. Not impossible, as shown by the large number of traditionally religious scientists, but difficult, as suggested by the dramatically lower rate of traditional religious belief among scientists as compared to the public at large.

As you might imagine, it is a source of frustration to me that most of my fellow Americans see things differently. I am not thinking now of political frustrations, such as the tendency of many religious folks to vote for candidates of whom I disapprove. That is annoying, yes, but right now I have something more fundamental in mind: I wonder what religious folks know that I do not. Do they have some insight I lack?

The issue is not simply that I have considered the philosophical arguments in support of God's existence and find them inadequate—though I have, and I do. It is that so much of religion is completely baffling to me. Not just fundamentalist religion, but also the more moderate form so often held up as a reasonable alternative. Since religion is a complex phenomenon with many aspects, let me clarify precisely what I find so puzzling.

At one level religion is about cultural identity. Participation in religious institutions often has more to do with socialization than it does with adherence to doctrine. Such participation is a way of announcing to others that you view yourself as part of that community. As I shall discuss in chapter 28, I find nothing baffling in *that* aspect of religion.

Others see religion as primarily about ritual and tradition. For such people, religion has more to do with behavior than with belief. That the rituals can seem odd to those looking in from outside is precisely the point. They have value and significance precisely because they are distinctive of a particular community. If the rituals could be defended on purely practical grounds then everyone would be doing them and they would lose their meaning. I can understand why someone might wish to preserve the rituals and traditions of their heritage, independent of whether they believe the underlying theological justifications.

For others the term "religious" suggests a certain attitude or disposition toward nature. It connotes a feeling of profound awe and smallness before the complexity of the world. Some might employ the language of holiness and divinity in describing such feelings, but it certainly is not necessary to do so. For example, biologist Richard Dawkins, one of the most outspoken atheists in the world, describes himself as "a deeply religious nonbeliever" (Dawkins 2006, ch. 1). Once again, nothing baffling here.

It is rather the adherence to doctrine that I find baffling. Whatever else it is, religion is also about putting forth a large collection of propositions relating to matters of empirical fact—many of which seem highly dubious. That the Earth is superintended by an all-powerful, all-knowing, all-loving God is already a tough sell given the ludicrous quantities of rottenness in the world, but that is the least of it. I have been told repeatedly, by people genuinely concerned about my soul, that Jesus was God in human form, lived a sinless life, and died on the Cross in payment for sins I would commit thousands of years later. They warn of the gruesome fate awaiting me in the afterlife, pending my failure to get with the program. In their more poetic moments they tell me I have a God-shaped hole in my heart that only Jesus can fill. I do not understand how people come to believe such remarkable things.

People keep telling me that the Bible has a significance far exceeding its historical and literary value, sometimes describing it with phrases like "the Word of God." It is unclear to me what reason there is for thinking it is any such thing. I might feel differently if the biblical stories were of such force and emotional power that we recognize in them eternal spiritual truths beyond the ken of any uninspired human writer. That there are many people who believe precisely that is hardly the cure for my bafflement. I can think of many literary works of demonstrably human origin that are far greater repositories of wisdom and moral insight.

Also mysterious to me is the attitude held by so many toward the intellectual status of religion. Among certain segments of the intelligentsia, it is common to describe science and religion as different ways of knowing, each valid in its proper domain. Reply to such sentiments with anything but a knowing nod and brace yourself for the stern looks you will receive. Persist in your skepticism, perhaps by asking what it is we can reasonably claim to know by applying such methods as religion provides, and you will quickly be accused of scientism, reductionism, intellectual arrogance, or some comparable epithet. But the question remains: What has religion actually given us to justify its status as a "way of knowing"? By what method do I distinguish correct theological assertions from incorrect ones? How is "bad theology" different from "theology I happen to dislike"?

Add to my bafflement the ease with which people speak of God. They seem to have no trouble at all believing that the God of traditional Christianity exists. They treat Him as a sort of explanation-for-free that resolves such mysteries as remain within the scientific picture of the world. In contrast to this, they regard the naturalistic view as so implausible as to be hardly worth bothering with.

For me, however, the whole idea of God gets stuck right at the start. It is not clear to me that it is even coherent to hypothesize an entity with the attributes God is said to have. There are philosophical difficulties, such as how to reconcile God's perfect foreknowledge with human free will,[1] but for the moment I have in mind other considerations.

"Mind without brain," for example, could well be a contradiction in terms for all the experience we have with actual minds. We are told that God exists outside of time and that to Him the past, present, and future exist simultaneously. It is possible to write those words, but can you really picture what they mean? I cannot, and to be perfectly blunt I don't think you can either. God is said to bring universes into being by acts of will. He is pure spirit, but can interact with matter in dramatic ways. He knows everything that everyone is thinking at every moment of every day. He exists necessarily, meaning that it is impossible for Him to fail to exist. Is any of this even possible? Is this really your easy-to-believe alternative to a naturalistic view of the universe? I wonder about other things as well. How does God keep from getting bored? With nothing that can challenge or surprise Him, how does He maintain His interest in much of anything? How does He deal with the loneliness of being the only one of His kind, or with the monotony of eternal existence?

Just to be clear: My point is not that my failures of imagination show that God does not exist. It is simply that invoking God to explain the universe seems like an instance of the cure being worse than the disease, or of filling one hole by digging others. To borrow an example from philosopher Robin Le Poidevin, it is like saying that smoke is coming from the house next door because some smoke-producing event is going on inside (Le Poidevin 1996, 37). I fail to see how our understanding is advanced by first hypothesizing God into existence, and then arbitrarily imbuing him with just the universe-producing properties we need Him to have.

From such experience as we have, it would seem that intelligence is rather limited in what it can bring about. It is natural forces that are endlessly surprising us with their richness and fecundity. In pondering the mysteries of existence we seem to have two options. We can hypothesize that the sorts of natural forces that have adequately explained about 99 percent of everything are also adequate for the few puzzles that remain. The alternative involves inventing an omnipotent deity, thereby saddling ourselves with a collection of conceptual difficulties far more disturbing than anything found in nature.

In the end I am not an atheist because I thereby have a ready answer to every existential question. Atheism just makes the fewest demands on my credulity.

Most of the time I am happy simply to disagree with theists, even the fundamentalists, and I am not hostile to religious people generally. It is nothing to me if other people want to live in accordance with various religious strictures and dogmas, so long as it is nothing to them that I choose not to participate.

There is one place, however, where the conversation turns far nastier. I am referring to the question of morality. It is commonly thought that those of us who lack faith in God are of dubious moral character. Most will grant that atheists can live moral lives, in much the same way that a dog can be trained to obey commands, but will wonder nonetheless whether we can really have any sound basis for our actions.

Theologian Keith Ward, for example, has some very definite ideas about what happens when nonbelief becomes too common:

> Those who wish to continue to think of the soul are concerned to protect the inherent preciousness of human life, its sacredness and possession of particular rights and obligations. Those who reject all such talk will see humans as objects for scientific manipulation; and the world they will come to inhabit, if it continues to exist at all, will be a world beyond morality, dignity and freedom; a world of the manipulation of truth, of morality and of ideology, in the name of whatever powers come to replace the God whom science has killed. (Ward 1992, 8–9)

Rabbi David Wolpe offers similar thoughts:

> The great tyrannies of the twentieth century were tyrannies of atheistic regimes: Mao's China, Stalin's Russia, and Hitler's Germany. Pol Pot's massacres in Cambodia were similarly done in the name of Marxist "liberation." When God is displaced from society, does the human ego expand proportionately to fill the absence? (Wolpe 2008, 67)

Later he writes:

> Living without God does not liberate the kindness trapped inside of us. The absence of God rather licenses the hubris and cruelty of people who cannot imagine something greater than themselves. The historical evidence thus far shows that the removal of God leaves a space for destructive ideologies, not benign humanism. (ibid., 71–72).

One would have thought it obvious that Hitler, Stalin, and Mao are not examples of what happens when a society freely loses its religious faith. They are instead the result of societies finding themselves in the grip of ruthless dictators with loyal militaries. The history of authoritarian theocratic regimes, in which religion is forced on a population against its collective will, is hardly inspiring, but Wolpe does not see in that an indictment of religious belief.

What is especially galling about Ward and Wolpe's armchair speculations is that the experiment has been done. Modern civilization provides several examples of societies in which free nonbelievers predominate. Not only do they provide no support for the horror stories of Ward and Wolpe, but they are actually among the most decent and socially conscious societies on Earth.

Sociologist Phil Zuckerman, summarizing his research into two of the world's least religious countries, writes,

> If there is an earthly heaven for secular folk, contemporary Denmark and Sweden may very well be it: quaint towns, inviting cities, beautiful forests, lonely beaches, healthy democracies, among the lowest levels of corruption in the world, excellent educational systems, innovative architecture, strong economies, well-supported arts, successful entrepreneurship, clean hospitals, delicious beer, free health care, maverick filmmaking, egalitarian social policies, sleek design, comfortable bike paths—and not much faith in God. (Zuckerman 2008, 2–3)

He summarizes his conclusions as follows:

> First of all, I argue that society without God is not only possible, but can be quite civil and pleasant. This admittedly polemical aspect of my book is aimed primarily at countering the claims of certain outspoken, conservative Christians who regularly argue that a society without God would be hell on earth: rampant with immorality, full of evil, and teeming with depravity. Well, it isn't. Denmark and Sweden are remarkably strong, safe, healthy, moral and prosperous societies. In fact, a good case could be made that they are among the "best" countries in the world, at least according to standard sociological measures. (ibid., 4)

Similar remarks could be made of several other European countries. Ward and Wolpe's arguments are not at all supported by the available evidence.

The question "Can we be good without God?" is one whose premise is so ludicrous and insulting that a proper response requires profanity. To admit it as a valid question is to grant a host of preconceptions that no one should be granting. There is no reason at all for giving theists a presumption of moral goodness, while placing atheists immediately on the defensive.[2]

Mind you, this sort of thing comes not just from fundamentalists, but from the highest echelons of religious thought. Nonbelievers are often lectured about respecting the depth and richness of theological writing, yet I routinely find in it the most casual contempt directed towards atheists. Respect, as I suggested earlier, is a two-way street. I will grant it to theology when the theologians seem inclined to reciprocate.

* * *

Which brings me to the New Atheists.

The past several years have seen the publication of several books strongly critical of religion. Most famous of these are Sam Harris's *The End of Faith*, Richard Dawkins's *The God Delusion*, and Christopher Hitchens's *God Is Not Great*, but many others have appeared as well. The extraordinary success of these books, in a country commonly thought to be entirely hostile to nonbelief, caught the eye of numerous media outlets. A 2006 article in *Wired* magazine (Wolf 2006) referred to these authors as the "New Atheists," and the name stuck.

It is not my purpose here to review the substance of these books. Rather, it is the response to them I find educational. Writing at the popular website Beliefnet, philosopher Michael Ruse described them as "violently anti-religion of all kinds" (Ruse 2009). Elsewhere he has written, "And this is why I think the New Atheists are a disaster, a danger to the wellbeing of America comparable to the Tea Party" (Ruse 2011).[3] Author Ron Rosenbaum, in an article for the online news magazine *Slate*, claimed that "some of them [the New Atheists] can behave as intolerantly to heretics who deviate from their unproven orthodoxy as the most unbending religious Inquisitor" (Rosenbaum 2010). Even the Dalai Lama joined in, writing in the *New York Times*,

> Though intolerance may be as old as religion itself, we still see vigorous signs of its virulence. In Europe, there are intense debates about newcomers wearing veils or wanting to erect minarets and episodes of violence against Muslim immigrants. Radical atheists issue blanket condemnations of those who hold to religious beliefs. In the Middle East, the flames of war are fanned by hatred of those who adhere to a different faith. (Gyatso 2010)

These are just a few items. There are many other examples.

In light of this risible hyperbole, it is easy to forget that all the New Atheists actually did was write a few books. Perhaps I should not be cavalier, though. After all, in these books, they were sometimes snide. They also speak publicly about religion, but only when invited to do so. Such activities hardly seem comparable to those of the most unbending religious Inquisitors, who were, let us recall, in the habit of torturing and imprisoning those with whom they disagreed. And since the day's headlines routinely bring reports of actual religious violence, I do have a basis for finding the New Atheists tame by comparison. This ludicrous overreaction illustrates, far more effectively than the books themselves, the excessive level of respect granted to religion in modern society. It is this knee-jerk respect, far more than the religious beliefs themselves, that I see as a serious social problem.

For all my skepticism and rational hard-headedness, the sheer number of smart, savvy people who take religion seriously gives me pause. I fear I do not find it as easy as some of my fellow atheists to be dismissive, though I certainly share their exasperation.

On several occasions, religious folks have asked me if I am "searching." In a sense I am. I am searching for an understanding of how so many of my fellow human beings see the world, because currently I find it incomprehensible. People I respect keep telling me there is more to religion than dubious supernaturalism, so I will keep looking. And who knows? Maybe someday the clouds will clear, the light will shine, and I, too, will know the joy of being smug towards nonbelievers. Until that happens I can only persist in my bafflement and marvel that religion continues to be the focus of so much emotional energy.

PART II

The Creation Mega-Conference, Lynchburg, Virginia, July 2005

6

An Introduction to Evolution

Let us begin by presenting those aspects of evolution most relevant to disputes with creationists.

Picture a population of your favorite animal, antelopes perhaps. The antelopes are not physically identical to one another, a fact partially explained by differences in their genes. We shall refer to the collection of genes found in a given antelope as the *genome* of that animal. The variant versions of a gene found at a given location in the genome are referred to as *alleles*.

Among the antelopes in our population, some will parent many children, some will parent very few children, and some will parent no children at all. To some extent their differing fortunes can be attributed to luck. A given antelope might be struck by lightning, for example, or be felled by some rare disease before reaching childbearing age. In other cases, their differing reproductive success can be traced back directly to their genes. Perhaps one antelope, because he possesses allele *X*, is able to run faster than antelopes lacking allele *X*. Or maybe allele *X* makes a male antelope more attractive than his competitors to female antelopes.

Imagine now that antelope by antelope we catalog each allele in each animal. We could then make a two-column table, with one column listing the alleles we find, and the other listing the percentage of antelopes in which each particular allele is found. The percentage of antelopes possessing a given allele is referred to as the *frequency* of the allele in that population. The collection of all of the alleles in all of the antelopes is referred to as the *gene pool*.

After the current generation of antelopes is finished having whatever offspring it is going to have, we could collect all of the offspring and make a new catalog of *their* alleles. We could continue making catalogs for generation after generation, line them up, and effectively watch as the composition of the gene pool changes over time.

As the generations go forth, some alleles, sadly, will go extinct. Other, more successful alleles might outcompete all of their rivals and appear in every antelope, a state of affairs referred to as *fixation*. Especially interesting is the appearance of novel alleles completely absent from the previous generation. Such novelties can arise in a variety of ways, but we shall abuse language somewhat by referring to all of them collectively as *mutations*. If we persist in our cataloging for a great many generations, we can imagine so many changes and so many novel mutations that our most recent catalog might bear almost no resemblance at all to the catalog with which we started. That is evolution. More precisely, evolution is the process through which allelic frequencies in gene pools change over time.

The process through which alleles conferring reproductively advantageous traits increase their frequency is called *natural selection*. It is an especially important mechanism of evolution, but it is hardly the only one. There is also *genetic drift*. A given antelope receives half of its genes from its mother and half from its father. Its genes are therefore a random sampling of the genes of its parents. This implies that a given allele might be present in a genome as the result of pure luck and not because of natural selection. We could identify other processes as well, but this is adequate for our purposes.

That evolution occurs, and that natural selection and genetic drift are two of the processes driving it, are observable facts of nature. These facts are accepted by creationists no less than by evolutionists.

The dispute arises when we pass from observable processes in the present to the broad sweep of natural history. The grand claim of evolution, rejected by creationists, is that the diversity of modern life is the result of natural processes like selection and drift operating over long periods of time, starting from an ancient population of unicellular, bacteria-like organisms. Evolution takes that ancient population for granted and makes no attempt to explain how life originated in the first place. The origin of life, while certainly overlapping and contiguous with evolution, is ultimately a different field of study.

In picturing evolution you should envision a family tree.[1] Just as my siblings and I share a recent common ancestor (our parents), and just as my first cousins and I share a more distant common ancestor (our grandparents), so, too, do any pair of species today share a common ancestor in the (perhaps distant) past. Animals that are anatomically very similar, like humans and chimpanzees, share a relatively recent common ancestor. Humans and dogs share a more distant ancestor, and humans and lobsters share a more distant connection still. It is important to keep in mind that in referring to a common ancestor of humans and lobsters, you should not be picturing something that is half-human, half-lobster. Given the enormous physical differences between humans and lobsters, it is safe to say that their common ancestor would not be at all recognizable as either one of them.

There is one aspect of evolution that seems to cause particular difficulty. Perhaps the most obvious characteristic of living organisms is their complexity. Viewed from an engineering perspective, the human eye and the wing of a bird, for example, are very impressive. They feature multiple parts interacting in such a way as to allow for a clear function. This is not the sort of thing one expects from natural causes alone. But evolution says there was a time when there was no such thing as eyes or wings. Then, much later, we have eyes and wings. Where did they come from?

Darwin himself understood the basic problem, but he also apprehended the solution. In *On the Origin of Species* he wrote:

> To suppose that the eye with all its inimitable contrivances for adjusting the focus to different distances, for admitting different amounts of light, and for the correction of spherical and chromatic aberration, could have been formed by natural selection, seems, I freely confess, absurd in the highest degree. When it was first said that the sun stood still and the world turned round, the common sense of mankind declared the doctrine false; but the old saying of *vox populi, vox dei,* as every philosopher knows, cannot be trusted in science. Reason tells me, that if numerous gradations from a simple and imperfect eye to one complex and perfect can be shown to exist, each grade being useful to its possessor, as is certainly the case; if further, the eye ever varies and the variations be inherited, as is likewise certainly the case; and if such variations should be useful to any animal under changing conditions of life, then the difficulty of believing that a perfect and complex eye could be formed by natural selection, though insuperable by our imagination, should not be considered as subversive of the theory. (Darwin 1859, ch. 6)

That is precisely the sort of answer modern biologists give. Complex structures form by a process of gradual accretion. We do not go in one generation from eyeless creatures to those with modern eyes. Instead we have a series of very small steps connecting the two states. Ascending a great height by using a flight of stairs provides an apt analogy.

Everyone who has contemplated the matter has been impressed by the intricate interworkings of living organisms. In describing structures like eyes and wings, it is difficult to avoid using words like "complex" and "designed." However, when you look more closely at the details, certain other phrases also become unavoidable. I mean phrases like "cobbled together" and "Rube Goldberg." These structures look like they arose gradually by an opportunistic process that had no foresight, not from the design of an intelligent engineer. In other words, they have just the structure they would need to have for natural selection to be a viable explanation.

7

An Introduction to Creationism

The Creation Mega-Conference took place on the campus of Liberty University in Lynchburg, Virginia, during one week in July 2005. Founded by the famed evangelist Jerry Falwell in 1971, Liberty is one of the few Christian colleges to promote young-Earth creationism.

The conference went from Monday to Friday, but registration began on Sunday night. I took advantage of this opportunity, hoping to avoid the inevitable lines the following morning. Alas, my plan came to naught. A long line of similarly inclined people snaked around the bare concrete floor of the Vines Center, the large sports arena serving as the main conference site (fig. 7.1). When my moment arrived, I swallowed hard, handed over my $150 registration fee and received my conference package: a complete schedule of presentations, a notebook, promotional material for various creationist groups, and a copy of the subtly titled book *The Lie: Evolution*, authored by Ken Ham. After transacting my business, I returned to the hotel and settled in for the evening.

Since the organizers decided Ham's book should be our formal introduction to creationism, perhaps we can use it for the same purpose.

Roughly speaking, creationism has two major facets[1]: As religion it is a view of the world premised on the inerrancy and perspicuity of the Bible on any subject it addresses. As science it is a collection of assertions meant to show both the inadequacy of modern evolutionary theory and the correctness of the creation account in Genesis. For most of the creationists I have met, the first facet takes precedence over the second. The basic attitude is aptly summarized by the slogan "God said it, I believe it, that settles it," which I heard repeatedly throughout the conference.

Ham's book is devoted entirely to the religious side of the issue. Its recurring theme is that Genesis is the foundation for all of the central tenets of Christianity. Discard what it says concerning the origin of the universe and of living things, and you are well on your way to moral relativism and anarchy. Ham writes:

FIGURE 7.1 The Vines Center at Liberty University.

> The biblical doctrine of origins, as contained in the Book of Genesis, is foundational to all other doctrines of Scripture. Refute or undermine in any way the biblical doctrine of origins, and the rest of the Bible is compromised. *Every single biblical doctrine of theology, directly or indirectly, ultimately has its basis in the Book of Genesis.*
>
> Therefore, if you do not have a believing understanding of that book, you cannot hope to attain full comprehension of what Christianity is all about. If we want to understand the meaning of anything, we must understand its origins—its basis.
>
> Genesis is the only book that provides an account of the origin of all the basic entities of life and the universe: the origin of life, of man, of government, of marriage, of culture, of nations, of death, of the chosen people, of sin, of diet and clothes, of the solar system . . . the list is almost endless. The meaning of all these things is dependent on their origin. In the same way, the meaning and purpose of the Christian gospel depends on the origin of the problem for which the Saviour's death was, and is, the solution. (Ham 1987, 59–61, italics and ellipsis in original)

Ham has no patience for people who would, in his view, twist the text of Genesis to accommodate "days" of indeterminate length or to allow for God's use of evolutionary mechanisms as his means of creation. Of course, there is no shortage of Christians who demur.[2] Ham's frustrated recognition of that fact is a common refrain. For example:

> These same warnings apply to many today who claim to be teachers of the Word of God, but who, in reality, are causing many people to fall by the wayside. Many of you will no doubt be aware that much of the opposition to the work of the creation organizations worldwide comes from theologians and other religious leaders. Many of the humanist groups often enlist people who claim to be Christians but believe in evolution to support them (on television, radio, and in publications) in their effort to combat the creation ministries. (ibid.,135)

For the past century and a half, evolution has been the dominant scientific paradigm. Virtually every university has either a Department of Ecology and Evolution or a significant division of its Department of Biology devoted exclusively to evolutionary studies. Peruse the journals in any decent science library, and you will see dozens devoted to evolution and its related fields. The library shelf space accorded to books on evolution is considerable, and new books are produced at a steady clip. Of course, none of this implies that the theory is correct, but it surely counts for *something*. It means that evolution is not a crazy, fringe idea. It is highly unlikely that a clever amateur will discover a fatal flaw that has been long overlooked by the professionals.

More than anything else it is the unwillingness of creationists to consider this point that I find distasteful. They seem to lack that small voice of skepticism telling them that things cannot really be as simple as their authorities pretend. How many times have I seen audiences cheer and applaud a speaker for leveling at scientists some preposterous calumny or absurd accusation? The implausibility of several generations of scientists being guilty of crass stupidity and gross incompetence never seems to occur to them.

This supreme level of confidence, I would say arrogance, is an immediate consequence of their religious views. Since they already have the Truth, rival ideas cannot be regarded as food for thought or as matters for debate. They are, instead, threats to be conquered. They are tools of "the Enemy," employed to shake the faith of weak Christians.

Ham writes, "My parents knew that evolution was wrong because it was obvious from Genesis that God had given us the details of the creation of the world" (ibid., 15). He elaborates later:

> After I gave a lecture to a class at a Christian college in Kansas, using material similar to that discussed already (plus additional scientific evidence), a student stated in front of the rest of the class, "What you have said sounds logical and very convincing in regard to accepting Genesis as truth. But, you must be wrong, because my geology professor here at the college believes in evolution and would totally disagree with you. If he were here now, I'm sure he could tell me where you are wrong, even if I can't see it at the moment."
>
> I replied, "Even if your geology professor were here and said things I don't understand because I'm not a geologist, if what he says disagrees with the Bible, then he is wrong. If I can't explain why he is wrong, it only means

> I don't have the evidence to know the errors in his arguments. The Bible is the Word of God and is infallible. I'm sure I could get a creationist geologist to find out why your professor is wrong, because the Bible will always be right!" (ibid., 48)

Chapter 8 of Ham's book is called "The Evils of Evolution," and it reveals another major aspect of creationist thinking. Over its ten pages we are treated to a gallery of terrible things laid at evolution's feet. These include lawlessness, homosexuality, Nazism, racism, drug use, abortion, social Darwinism, and male chauvinism. The theme underlying this charming little list is that accepting creation implies accepting God's sovereignty over your life, which includes adhering to His laws and standards. Evolution is equated with a rejection of these principles and therefore with placing man's judgment ahead of God. This leads to an abandonment of God's principles and rampant immorality.

We could certainly go item by item and find much to criticize. The casual inclusion of homosexuals on a list that includes Nazis, racists, and drug users is rather obnoxious. We could agree to the general premise that various unpleasant people have twisted Darwin's ideas to justify their rottenness without agreeing that this has the slightest relevance to evolution's correctness as a scientific theory. Christianity, after all, also has a long history of being used to justify evil.

I would prefer instead to make a different point. You might think it is a blow to Ham's argument that human immorality long predates Darwin. You think this, no doubt, because you also think of evolution as a scientific theory pioneered by Charles Darwin in 1859. Ham and his supporters use the term somewhat differently. For them "evolution" is simply synonymous with "anti-God." To say that someone accepts evolutionary thinking is to say that he rejects God's laws and Word and commandments. Seen in this way, Darwin's research and those of modern scientists following his lead are simply a manifestation of evolution. They are modern representatives of a view that has existed for almost as long as there have been human beings at all.

For scientists trying to engage creationism, this all seems bizarre. It looks like a delusion so convoluted that it is unclear even where to find a point of entry for correction. To those Christians who long ago made their peace with evolution, it will come as news that they have actually rejected God. Scientists think of evolution as a useful theory, not as an all-encompassing worldview. The modern theory of evolution certainly has consequences for many areas of human thought, but it is simply a factual error to suggest a necessary connection between evolution and any other political, religious, or ethical view.

I conclude with one more quotation. Since this was the last thing I read before going to sleep, you can imagine my dreams that night.

> Dear reader, there is a war raging. We are soldiers of the King. It is our responsibility to be out there fighting for the King of kings and Lord of lords. We are the King's army. But are we using the right weapons? Are we fighting the battle where it really matters? Unfortunately, many Christians have what would be viewed militarily as a totally ridiculous strategy. They do not fight the battle where it rages. They are not fighting on the real battleground. They have no hope of winning. When are Christians in the nations around our world going to wake up to the fact that we need to re-aim our weapons and aggressively and actively fight the issue of evolution by restoring the foundation of creation? (ibid., 109–111)

8

Literalism and Other Canards

When we speak of anti-evolutionists, just how many people are we talking about?

Since 1982 the Gallup Organization has, at regular intervals, asked people in the United States this question:

> Which of the following statements comes closest to your views on the origin and development of human beings?
>
> 1. Human beings have developed over millions of years from less advanced forms of life, but God guided this process.
> 2. Human beings have developed over millions of years from less advanced forms of life, but God had no part in this process.
> 3. God created human beings pretty much in their present form at one time within the last 10,000 years or so.

For simplicity, let us refer to these options, respectively, as theistic evolution, non-theistic evolution, and creationism. Their findings for a few representative years are shown in table 8.1. The data shows roughly stable percentages for each of the three options over the life of the poll. It would seem that just under one-half of the American population supports creationism.

However, it would be very misleading to end the discussion there. Public polling in this area is fraught with peril. People's religious views are generally too complex to capture with simple questions. Social science research has found consistently that small differences in question phrasing can translate into significant changes in the results (Berkman and Plutzer 2010, 32–63). Other polling has shown that Americans are largely ignorant of basic facts relevant to this question. The implication is that people's answers are based as much on emotional responses to key phrases in the question as they are on careful consideration of the relevant facts and evidence.

This suspicion is borne out by more careful polling. Political scientists George Bishop, Randall Thomas, Jason Wood, and Misook Gwon polled people on a variety of specific assertions related to evolution and creationism. They found that a mere 18 percent of people agreed with the statement that "the earth is less than ten thousand years old." Seventy-eight percent agreed with the statement that "layers

TABLE 8.1

Public Acceptance of Evolution in the United States, by Year

Year	Theistic Evolution	Nontheistic Evolution	Creationism
2010	38	16	40
2007	38	14	44
2004	38	13	45
2001	37	12	45
1997	39	10	44
1993	35	11	47
1982	38	9	44

Source: Gallup (2010)

TABLE 8.2

Percentage of U.S. Population, among Different Religious Groups, Agreeing or Disagreeing That Evolution Is the Best Explanation for the Origins of Human Life

Group	Agree	Disagree	% of Population
Evangelical Protestant	23	70	26.3
Mainline Protestant	51	42	18.1
Historically Black Churches	39	51	6.9
Catholic	58	35	23.9
Orthodox	55	36	0.6
Jewish	77	17	1.7
Muslim	45	51	0.6
Buddhist	81	14	0.7
Hindu	80	14	0.4
Mormon	21	76	1.7
Atheist	87	9	1.6
Agnostic	87	11	2.4
Jehovah's Witness	8	90	0.7

Source: Pew Forum on Religion and Public Life (2008, 95)

of rock containing fossils cover the earth's surface and date back hundreds of millions of years." They also found quite a few instances of conflicting results. Thus, despite the small level of acceptance for a young Earth, fully 50 percent of respondents agreed with the statement that "the Bible describes the creation of life exactly as it occurred in six days," and 60 percent agreed with the statement that "all people are descendants of one man and one woman—Adam and Eve" (Bishop et al. 2010).

Also relevant here is data from the U.S. Religious Landscape Survey carried out by the Pew Forum on Religion in Public Life. They asked people if they agreed or disagreed with the statement that evolution was the best explanation for the origins of human life. Their results, categorized by religious affiliation, are presented in table 8.2. We have also included the percentage of the U.S. population represented by each affiliation.

Political scientists Michael Berkman and Eric Plutzer have carried out the most detailed analysis of the public opinion polls to date. Let me close this section by quoting from their work:

> We uncovered several possible sources of bias—in which question wording or the theme of the survey could change the level of support for creationism by five or even ten percentage points. However, we also saw substantial evidence of consistency and evidence that deliberate attempts to manipulate answers had meager effects. . . .
>
> Even if the actual percentages might differ from poll to poll, there can be no doubt that the large majority of Americans want creationism taught in science classes. A large plurality of this group wants creationism taught *as science and in science classes*, whereas others apparently would be satisfied to see some ideas of creationism validated by their discussion as a religious perspective or belief. (Berkman and Plutzer 2010, 62–63)

There were several chartered buses outside the Vines Center when I arrived for the Creation Mega-Conference at 7:30 on Monday morning. To judge from the license plates, some of my fellow conference-goers had come a very long way. That impression was confirmed as I took my seat in anticipation of the opening presentations. I struck up a conversation with the older couple sitting next to me, who looked to be in their seventies. I asked them where they were from, and they told me they had come from Florida with a church group. I raised an eyebrow and asked them why they had gone to so much trouble. They replied that they had long admired Jerry Falwell; since the timing worked out, they were eager to see him in person.

Scanning the audience I saw that senior citizens were well represented, as were families with young children. Minorities, by contrast, were all but nonexistent. The first speaker was Falwell himself, to the evident gratification of my conversation partners. There was raucous applause when he took the stage. His speech consisted entirely of standard fundamentalist boilerplate. Since little will be gained from a detailed consideration of his remarks, let us use this time to clear up a few common misconceptions about creationists.

They are not biblical literalists, at least not in the very crude sense their critics typically envision. They understand, along with everyone else, that the Bible includes many literary genres. They argue, however, that if you are going to interpret a passage nonliterally there should be strong textual grounds for doing so. They sometimes say, “If the plain sense makes sense, then seek no other sense.” If a verse’s intent is poetic or symbolic, that should be clear from the context and the writing style. They believe strongly in the “perspicuity of scripture.” This is a foundational principle of Protestant theology, asserting that the major teachings of scripture should be comprehensible to a person of normal intelligence, reading the Bible in his or her native language. Certainly there will be obscure passages and translational ambiguities, but those portions of the Bible essential to understanding the grand story of creation, fall, and redemption should be clear to all. If, for all that, a passage is unclear, then we should turn to other, more transparent parts of scripture for clues. This exegetical principle is sometimes described by the pithy phrase, "Let scripture interpret scripture."

For example, in Genesis 4:8–10 we find:

> Cain said to his brother Abel, "Let us go out to the field." And when they were in the field, Cain rose up against his brother Abel and killed him. Then the Lord said to Cain, "Where is your brother Abel?" He said, "I do not know; am I my brother's keeper?" And the Lord said, "What have you done? Listen; your brother's blood is crying out to me from the ground!"

Taken literally, this would suggest that prior to speaking with Cain God did not know of Abel's murder. Since this contradicts God's omniscience, we have sound reasons for favoring a nonliteral interpretation. God does not need to converse with humans to learn facts about the world. Creationists interpret these verses as condescensions by God to limited human comprehension. The conversation between God and Cain is a literary device for showing us that, despite Cain's feeble attempt at concealing his crime, our sins cannot be hidden from God.

Here is a second example, this time from the New Testament, Mark 4:30–32:

> [Jesus] also said, "With what can we compare the kingdom of God, or what parable will we use for it? It is like a mustard seed, which, when sown upon the ground, is the smallest of all the seeds on earth; yet when it is sown it grows up and becomes the greatest of all shrubs, and puts forth large branches, so that the birds of the air can make nests in its shade."

It turns out that the mustard seed is not the smallest on earth. There are certain orchids with smaller seeds. It would seem, then, that Jesus has said something that is false. Surely, though, we are on solid ground in thinking that Jesus was not trying to teach botany. The mustard seed was the smallest of which the ancients would have been aware, so Jesus employed an exaggeration to make a spiritual point.

Now contrast this situation with the first chapter of Genesis. The reference to the mustard seed is clearly identified as an analogy and a parable, and the religious lesson being taught is self-evident. In these points the story differs from Genesis 1, which reads like a straightforward historical narrative. There is nothing in the Genesis creation account to label it as a parable, or an allegory, or an accommodation to limited human understanding. Creationists bolster this claim with textual analyses meant to show that the language of the original Hebrew is consistent with passages of the Bible that everyone acknowledges are historical, and are not consistent with passages meant figuratively (Beall 2008, Boyd 2008)[1]

The argument continues by considering the specific events related in the text. The story of man's creation as the pinnacle of God's work, especially when combined with the events of the following two chapters, are seen as absolutely central to the entire Christian story. They are not small details presented in the service of a greater point. In the New Testament, Adam is treated as an historical figure, and his sin is specifically identified as the reason humanity needs a savior at all. How can you turn the Genesis stories into an allegory, creationists ask, when these events are relevant to so much that comes later?

Now ask yourself how it looks to them to suggest that modern science compels us to an allegorical interpretation of Genesis. Not only are you placing fallible human reason ahead of infallible scripture, but also you are telling them they need a team of biologists, classicists, and literary theorists to interpret the Bible for them. Not, mind you, regarding a few isolated verses addressing obscure historical questions, but with regard to the foundational stories of the entire volume.

We come next to the tiresome assertion that YECs err in thinking Genesis is a science textbook. They don't—at least, again, not in the simplistic way that is often suggested.

When the sixty-six books of the Bible are published in a single volume, the result is a long, dense tome almost none of which has anything to do with science. The creationists are as aware of this as anyone else. Like all traditional Christians, they believe the Bible's primary purpose lies in the telling of the grand narrative of human history. It instructs us regarding our need for and the availability of redemption.

Such purposes extend to the early chapters of Genesis. The six-day creation account is not primarily a lesson in cosmology or biology. As creationists see it, its major goals are theological. These passages establish that God created everything we see in nature. This was a rebuke to the pantheistic religions common to the place and time of the text's compostion—the Middle East of the fifth and sixth centuries BCE—which regarded the sun and the moon, for example, as objects of veneration. The viewpoint of the authors of Genesis 1 is unabashedly monotheistic, which was in contrast to the common polytheism of their time. The text establishes the initial perfection of God's creation, later sullied by human sin. By creating in six days and resting on the seventh, God provided a model for us to follow. Establishing these truths are the primary purposes of the text, argue the creationists, and none of them involve instruction in science.

Creationists do, however, believe that the Bible is inerrant on any subject it addresses. If that means accepting what it says during its very rare excursions into science, then so be it. In laying out the basic truths of the human condition, the biblical scribes expressed themselves in ways that have scientific consequences. We might find it interesting and suggestive that the Bible, which mostly avoids scientific questions, opens with so much of a scientific nature. Perhaps the conclusion is that God considered these particular scientific truths to be so important that they could not be omitted without compromising the story. To the person who says, "Genesis was never intended to teach us science," creationists reply, "Then why does it contain so much of scientific relevance?"

The basic view is well summarized by Henry Morris:

> It is obvious, of course, that the Bible is not a scientific textbook in the sense of giving detailed technical descriptions and mathematical formulations of natural phenomena. But this is not adequate reason for questioning the objective accuracy of those numerous portions of Scripture which do deal with natural phenomena and historical events.[2] (Morris 1964)

Let us also be clear that creationists do not make an acceptance of their view of Genesis essential to one's status as a Christian. Your salvation does not hinge on thinking the days of Genesis are of the normal, twenty-four hour sort. They share with all mainstream Christian denominations the belief that salvation is a free and undeserved gift from God, available to anyone who acknowledges both his or her hopelessly sinful state and the sacrifice Jesus made for him.

Here is Duane Gish, a prominent YEC, making the central point:

> Christians may differ on the interpretation of certain doctrines, such as: Does baptism require immersion, and is it reserved for those of responsible age? Are the bread and wine of the Lord's Supper transformed into flesh and blood? Is the millennial rule of Christ a literal thousand-year reign on this earth, yet to come? Does the Hebrew word *yom* in Genesis 1 denote a solar day, or may it refer to a long period of time? By what method did God create the heavens and the earth? These are important matters, and they have occupied theologians for centuries. But should one's salvation depend upon getting all the answers right? Must that be done to "believe on Christ"?
>
> No, not at all! God did not make salvation difficult to understand and to obtain. Our faith is simple, like a child's faith in a father.[3] (Gish 1989)

Why, then, the obsessive focus on Genesis? There are two reasons: The first is that creationists genuinely believe that theirs is the only interpretation of the Bible that makes sense. As they see it, attempts to treat the creation account nonliterally do violence to the text and are contrary to its plain teaching. This leads to the second reason, which is the slippery slope they perceive leading from a nonliteral interpretation of Genesis to wrong-thinking on more central issues. Allowing Genesis to make room for evolution is a "compromise" made to accommodate the Bible to earthbound human wisdom. Having established your willingness to compromise to earn favor with secular science, what will you do when the authorities tell you the Virgin Birth and the Resurrection are also scientifically implausible?

9

Browsing the Bookstore

Back at the Vines Center, after Falwell's initial speech, the next major talk came from Ken Ham himself and was entitled "Rebuilding the Foundation." It was mostly a cheerleading talk, with little scientific content. His frequently repeated rallying cry was "We're taking them back!" The idea was that secularists and atheists had corrupted many areas of culture and science, but the true Christians were now coming to reclaim them. Here is a partial list of things Ham described as requiring retrieval by the Christian community: Christian institutions, history, animal kinds, biology, genetics, DNA, the meaning of death, physics, geology, the Grand Canyon, stalagmites, stalactites, dinosaurs, marriage, "people groups," and education.

You can probably guess their logic: All too many Christian institutions have fallen into the hands of compromisers willing to accept evolution. A proper study of biology must be premised on the concepts of Genesis 1. Geology took a disastrous wrong turn when it rejected the realities of a young Earth and global flood. Public education is entirely in the hands of secular humanists.

Ham ended with an exhortation to spend lots of money in the bookstore. Since we now had a thirty-minute break before the next talk, I decided to take a look.

Flipping through books at random, you quickly notice a lingo to young-Earth creationism that takes some getting used to. The word "compromise" is ubiquitous. It is usually a term of abuse hurled at Christians willing to accept a nonliteral interpretation of Genesis. The term "science" is often modified with unusual adjectives. For example, there is "secular science," which refers to science as it is practiced outside of the creationist subculture. This is contrasted unfavorably with "true science" or "Bible-based science." A phrase like "millions of years" (occasionally, "billions of years") is used as short-hand for the notion that the Earth is much older than a few millennia. The Bible is often referred to as "scripture" or the "Word of God," this last sometimes shortened to "the Word" or "God's Word." Thus, you might read something like, "Some Christians have compromised with secular ideas like millions of years. But the Word says the Earth is only thousands of years old, and true science confirms this."

FIGURE 9.1 A display of creationist books.

There are other adjectives placed before "science." An example is "origins science," which is a blanket term referring to any discipline relevant to understanding how human beings came to populate the planet. There is also "historical science," which tries to reconstruct natural history, as contrasted with the real thing, which is referred to as "experimental," "observational," "operational," or "empirical science." These are hugely different enterprises. No one can directly observe the past, you see. We can only view the facts of the present through the lens of our biases and preconceptions—that is, through our "worldview" or "belief system." This idea is captured in the creationist slogan, "Same facts, different starting points."[1] Of course, creationists view their starting point as vastly superior to that of secular science. They believe that "God decides truth." Evolutionists, by contrast and to their detriment, believe "man decides truth."

Genesis 1 makes frequent reference to the different "kinds" of plants and animals. To creationists this is a meaningful term in biological classification, much like the terms "species," "genus," and "family," which are so familiar from mainstream science. "Kinds" are not easy to define; they do not correspond perfectly to the idea of a species. In some cases the technical term "baramin," from the Hebrew words *bara* (created) and *min* (kind), is used. The systematic investigation of baramins is known as "baraminology." This should be viewed as a proper branch of science, complete with its own methodology, conferences, and technical publications.

Nor is baraminology the only discipline that creationists claim to have pioneered. There is also "flood geology," in which the reality of Noah's flood is taken as

FIGURE 9.2 This display at the Creation Museum (see part four) shows the wrecking ball of "Millions of Years" destroying the foundation of a church.

a central datum for geological practice. In its modern form it was inaugurated by Henry Morris and John Whitcomb Jr. in their 1961 book *The Genesis Flood.* One of flood geology's proudest accomplishments to date is the theory of "catastrophic plate tectonics," which attributes the motion of the continents to the violent activity of the flood. Of course, the totality of the models and theories creationists have developed are referred to collectively as "creation science" or "scientific creationism." Creation science takes a strong interest in the days enumerated in Genesis, which are referred to collectively as the "Creation Week."

Creationists argue that evolutionary change can occur, but only within the confines of the originally created kinds. You will therefore see a distinction between "microevolution" and "macroevolution." The former refers to evolution within a kind, which is an empirical fact to which creationists do not object. The latter refers to evolution on a larger scale, which has never been observed and is just a dubious conclusion drawn by secular scientists engaged in historical science based on a materialist worldview.[2] Creationists wishing to mock the idea of large-scale evolution will often refer snidely to evolution from "molecules to man," "particles to people," or "fish to philosopher." Sometimes "microbes to man" is the favored formulation. If you prefer assonance to alliteration, there is also "from goo to you by way of the zoo."

It is not uncommon for familiar terms to take on new connotations. "Assumptions" are very bad. Typical usage: "Secular scientists say the Earth is very old because of radiometric dating, but that method is based on all kinds of *assumptions.*"

"Mutations" are also very bad. They are nearly always harmful to the organism and *always* lead to a loss of "information." That last term is never adequately defined, and usually has little relation to the various technical definitions used by mathematicians. In creationist usage it refers vaguely to something that conveys a meaningful message. More importantly, it is inherently nonphysical. This is said to be a blow to materialist conceptions of the world.

"Evolution" refers not to a well-supported scientific theory, but to an all-encompassing, God-denying worldview closely associated with secular humanism. "Theory" is a term of abuse, as in "Evolution is *just a theory*." It is not uncommon for those last two words to be combined into the awkward phrase "evolution theory," as in, "The origin of new genetic information is a problem for evolution theory, but not for creation science." Also common is the term "evolutionism," to emphasize its status as an ideology.

Plurals appear where you do not expect them. Creationists speak of "evidences" and "proofs," this latter term being borrowed from Acts 1:3, "After his suffering he presented himself alive to them by many convincing proofs, appearing to them over the course of forty days and speaking about the kingdom of God." Religious terminology is sometimes used in scientific contexts. Scientists are said to "believe in" evolution, for example. Evolution "teaches" that all species are related by common descent.

Young-Earth creationism is mostly a subculture of Christian evangelicalism, and it therefore inherits many evangelical figures of speech. Young-Earth creationists believe that God has "a plan for your life"—by which they mean something God put you on Earth to do. You learn of this plan by heeding God's "call." Someone might say, "I felt called to the ministry," or "I felt called to pursue graduate studies in science." If you feel adrift and unhappy in your current circumstances, it could mean you failed to heed God's call toward His plan for your life. Whereas God's call usually refers to the broad themes of your life, it is God in the form of the Holy Spirit that is more active in your day-to-day activities. Creationists will often say they felt "the Holy Spirit leading them" toward a particular course of action.

The word "ministry," incidentally, refers to any sort of activity meant to bring souls to Christ. For example, if your activities are focused on the issue of evolution and creationism, you might say you are engaged in "Creation Ministry."

Non-Christians are referred to as "lost," "unsaved," "unchurched," or "dead in sin." Someone making a sincere effort to learn the Truth about Jesus Christ might be said to be "searching," or "seeking." On the other hand, if you have had a conversion experience, then you are "saved," or "born again." You have "come to Christ," and now have a "saving faith." You have started your "Christian walk." You have "come to know the Lord." This is a moment of tremendous power in the life of a believer, and it is one they are happy to tell you about with the slightest encouragement. When a believer tells you the story of his conversion, he is said to "give his testimony." During the time it takes for him to tell his story he is "witnessing" to you.

For Evangelicals, conversion represents a liberation from the depravity of "the world," or "the culture," by which they mean the world outside of their own subculture. The world is frequently described as "fallen," or is said to be "groaning under the curse"—that is, fallen from God's originally created perfection as the result of Adam and Eve's sin, and groaning under the curse God placed on the world as a result of that sin. Roaming freely through the world is Satan, who is often referred to simply as "the Enemy." Evangelicals are commanded to be "in the world, not of the world."

Creationist discussions of public education bring an additional lexicon to master. Creationists see "two models" for origins, one based on "creation," the other based on "evolution." These should be regarded as blanket terms, meaning roughly that you either invoke God in explaining the origin of human beings (creation) or you do not (evolution). Prior to several adverse court decisions in the 1980s, creationists called for "equal time" laws to be passed, requiring schools to give equal time to the creation model whenever the evolution model was discussed. Nowadays the usual catchphrase is "teach the controversy" over evolution and creation, though this phrase is more common among advocates of intelligent design. Scientists reply, quite correctly, that such controversy arises only to the extent that it is manufactured by religious demagogues trying to evade the separation of church and state.

Since they have been largely unsuccessful in getting creationism formally adopted in public schools, creationists are left instead to rant and froth. Public schools are "indoctrinating" students in the "godless philosophy of secular humanism." When extra derision seems called for, they refer to "government schools."

As I said, it all takes a bit of getting used to.

10

The Best Evidence That God Created

The Conference now featured parallel sessions, one described as "basic," the other as "advanced." I chose the basic track, which featured "What Is the Best Evidence That God Created?" by Carl Kerby.[1]

Kerby, a large gentleman with a voice made for public speaking, got off to a good start by telling us this would be a fun talk. He was certainly enthusiastic enough. He presented a sampling of nature's oddities, emphasized their beauty and intricacy, and then asserted bluntly that they could not be the products of gradual evolution.

The tone was established early. Kerby showed a slide of an intricate sand sculpture on a beach. He said, "You tell me how millions of years of random chance processes is going to cause something like *that* to happen. I tell you the heavens declare the glory of God." That last is a reference to Psalms 19:1.

Kerby now launched into his menagerie of evolution-defying critters. He showed us the cave weta, an insect native to New Zealand. One particular species of weta is found in caves at the top of mountains. In winter the temperature drops to well below freezing. The weta freezes into a state of suspended animation for months at a time. When the spring comes, the insects thaw and go on their way.

Next up was the moloch lizard of Australia. These particular lizards are desert animals, meaning the search for water is a major feature of their existence. They absorb water through a system of capillaries in their skin and also possess a small structure near their throats that serves as a pumping mechanism. In this way they can take advantage of minute amounts of water, such as you might find in wet sand.

The list continued. We were shown the mating behaviors of the emperor penguin; the remarkable transformative abilities of the mimic octopus; a gallery of frogs possessing various God-glorifying habits and physical attributes; and, of course, human beings—especially our eyes and our processes of embryological development.

After each item Kerby asked us, "Is this the best evidence that God created?" Prompted by a slide the audience was delighted to reply, "Not even close!"

The best evidence that God created, you see, is that the Bible says He did. Kerby said:

> Too many of us are looking for that evidence to prove the Bible—it doesn't work that way. You use the Bible to understand the evidence. You use the Bible to understand the world that we live in. Because if you want to argue about the fish and the birds and the rocks; every time they find a new bird, a new fish and a new rock you're in an argument again. When you start from the one who's always been there, who told you what happened, you can know that God created.

As presented by Kerby, it sounded like a bad thing that the discovery of new birds, fish, or rocks leads us right back into an argument. This exemplified an attitude toward nature fundamentally different from how scientists generally see things. Presented with new data, a scientist sees a welcome opportunity to update and revise the present best models of nature. Gathering new birds, fish, and rocks (along with other sorts of scientific data, of course), and then using them to refine our understanding of nature's workings, is precisely the point of science.

That is why nothing in science is ever put forth as the last word. Instead it is always, "This is the best we can do, for now." Science does not deal with absolute, capital *T* Truth. The very concept is foreign to proper scientific discourse. We deal instead with degrees of confidence. Our best theories account for so much data and have proven themselves so successful in practical situations that it would just seem perverse not to describe them as "true." Even then, however, we must preserve that small, nagging voice of skepticism.

More troubling than Kerby's elevation of the Bible to a status above that of the physical evidence was his toxic emphasis on certainty. Even suggesting the concept of an infallible source of information about nature entails the abandonment of the scientific method.

Something else struck me. At no point did he present an actual argument to show that the items in his little gallery were beyond what evolution could produce. We were given instead a list of curiosities, each followed by an assertion that evolution had been refuted.

It has never been easier for an inquisitive person to learn the state of play in any field of inquiry. If you are going to address an audience on subtle questions of science, especially if your thesis is that scientists are in thrall to a blinkered, inadequate understanding of their disciplines, it behooves you to spend a few minutes with Google. As preparation for writing this section, I resolved to see what I could find.

I began with the cave weta. It was the work of a few minutes to find a scholarly article called, "Evolution of the Diverse Antifreeze Proteins," published in 1998. The paper's abstract (academese for "summary") was very suggestive:

> Different types of ice-growth-inhibiting antifreeze proteins, first recognized in fish, have now been isolated from insects and plants, and the list continues

> to expand. Their structures are amazingly diverse; how they attain the same function are subjects of intense research. Evolutionary precursors of several members have been identified—divergent proteins of apparently unrelated function. The hybridization of information from structural and molecular evolution provides a forum in which issues of selection, gene genealogy, adaptive evolution, and invention of a novel function can be coherently addressed. (Cheng 1998, 715)

I found many other papers along these lines. In some cases meticulous genetic and molecular comparisons allowed researchers to identify evolutionary precursors for specific proteins. More generally, working within a paradigm of evolution permitted gradual progress to be made. The literature regarding the evolution of antifreeze proteins is enormous, amounting to hundreds of papers.

Insects are not the only animals to possess antifreeze proteins. There are also various species of Antarctic fish to consider, and the literature records substantial progress in understanding the evolution of these proteins as well. Summarizing recent research, cell biologist Kenneth Miller writes:

> One of the most dramatic examples of evolution using old genes for new purposes is that of Antarctic fish. The southern polar ocean began to freeze over a little more than 10 million years ago, producing a strong selective pressure in favor of fish that might be able to keep their blood from freezing in the increasingly chilly waters. As a result a variety of fish underwent evolutionary adaptations that modified preexisting genes to produce proteins that served as a kind of biological anti-freeze. One of these has been studied in such detail that we now know exactly where it came from. Evolution didn't merely take an existing gene and cobble a few changes into it. Rather, it started with an extra copy of a gene for a digestive enzyme. Mutations deleted most of the original gene, and then duplicated, over and over again (forty-one times, to be exact), a small portion at the beginning of the gene. The modified gene was then expressed in the liver, and the protein it codes for is targeted to the bloodstream. The duplicated region works as a powerful antifreeze agent, and the fish thrive in the chilly waters. The new gene is so different from the original that it might be difficult to recognize were it not for the fact that both ends of the antifreeze gene match the "old" gene of the digestive enzyme, betraying its origins and marking it clearly as the product of rapid evolution. (Miller 2008, 85, summarizing Logsdon and Doolittle 1997)

My findings were similar in the case of the Moloch lizard. I had little trouble finding articles discussing the mechanisms by which desert lizards absorb water from their surroundings, but none of them argued that they embodied any particular challenge to evolution. Instead the findings were typical of what we would expect: Desert lizards use a variety of mechanisms for extracting water, but they all appear to be variations on a basic theme (Sherbrooke et al. 2007). This is precisely the sort of pattern typical of evolution by natural selection. Focusing on lizards generally, as opposed to their water-extraction mechanisms specifically, I found whole books discussing their evolution (Pianka and Vitt 2003). If evolutionary biologists feel defeated by these critters, I could find no evidence of it in my research.

I have no doubt the creationists would respond by accusing scientists of spinning tales. The charge is that they simply assume evolution and then tailor their hypotheses accordingly. This sort of criticism is common among creationists, but it is very unfair. Worse, creationists tend to be inconsistent. First, they point to some complex adaptation and loudly proclaim it absurd to think it evolved gradually. Then, when scientists dutifully uncover likely precursors and plausible gradualistic scenarios, they say it is trivial to make up a story.

Scientists do not just make up evolutionary stories. Instead they begin by doing the hard work of understanding, in meticulous detail, how the adaptation in question actually works. This is often a difficult problem all by itself. By comparing similar structures in other organisms, they then search for scenarios to account for their evolution. Such scenarios must be consistent with all that is known, a heavy evidential burden. Having devised a plausible scenario, scientists do not then regard the problem as solved. Instead they use the scenario as the starting point of further research. In this way evolutionary thinking drives the discovery of novel facts about nature.

The point is not that scientists can give a mutation-by-mutation account for the origin of the weta's antifreeze proteins or the water-capturing mechanisms of desert lizards. Rather, the point is that it is the creationists who make bold, unsupported claims about what is possible and what is not. It is for them to back up their claims with arguments and evidence. The scientists who actually study these systems, the ones who know everything there is to know about them, do not believe they are seeing anything disconcerting to evolution. They routinely use evolution to drive their research and are met with tangible progress for their efforts.

If the creationists want them to give up all of that, they will need something more compelling than evidence-free expressions of disbelief.

On several occasions creationists have asked me, "What is the best evidence for evolution?" Typically this is posed as a challenge, as though they are prepared to refute whatever item I might produce. After Kerby's talk I gave the question some serious thought.

The proper response is that the question is misguided. Our confidence in evolution is not based on just one line of evidence, but from its ability to account for the data in many diverse lines of inquiry. It misses the point to focus on any one branch of thought.

Alas, this answer, while entirely correct, is rhetorically weak. It makes you look evasive in debate.

Personally, I have always found the fossil record compelling. Each fossil is a snapshot telling us that an animal possessing certain anatomical features existed at some point in our distant past. When these fossils are arranged in chronological order, we notice two things.[2] With literally millions of fossils to examine, not a single one is out of place from an evolutionary standpoint. The simplest fossils appear at the beginning, with more complex organisms appearing only after simpler precursors have paved the way. Looking more closely, we find numerous instances of transitional forms bridging the gaps between major groups (Prothero 2007). The perfect consonance between physical data and theoretical expectations is mighty coincidental if the theory is wrong.

Here is another line of evidence I find striking. It sometimes happens that viruses invade a larger organism and leave copies of their genomes embedded in the DNA of their host. These embedded genomes are referred to as "endogenous retroviruses," and modern geneticists have identified numerous examples. Curiously, they have found no less than seven cases where humans and chimpanzees share the same retrovirus at precisely the same place in their genomes (Theobald 2007). The location at which the virus embeds its genome is essentially random. It is asking a lot from chance to think that on seven different occasions viruses invaded both humans and chimps and left their remains in precisely the same location in each. The simpler hypothesis is that the virus invaded the common ancestor of humans and chimps, which then passed on those remains to all of its descendants.

This evidence is certainly suggestive, and I could continue in this vein by discussing embryology, biogeography, molecular biology, and numerous other fields besides. But if you ask me bluntly whether any one of these represents the best evidence of evolution, I am tempted to reply, "Not even close!" The best evidence, I would think, is simply that evolution works. When scientists apply evolutionary thinking to the practical problems they face, they are routinely rewarded with tangible progress.

For example, a few years back a team of paleontologists led by Neil Shubin became interested in the evolutionary transition from fish to land-dwelling animals. The extant fossil evidence suggested that this transition occurred between 365 and 380 million years ago. They looked for sites at which exposed rocks of that age could be found, ultimately settling on a remote location in the Arctic. They subsequently found a remarkable transitional form. Shubin writes:

> But our new creature broke down the distinction between these two different kinds of animal. Like a fish, it has scales on its back and fins with fin webbing. But, like early land-dwelling animals, it has a flat head and a neck. And, when we look inside the fin, we see bones that correspond to the upper arm, the forearm, even parts of the wrist. The joints are there, too: this is a fish with a shoulder, elbow and wrist joints. All inside a fin with webbing. . . .
>
> It took us six years to find it, but this fossil confirmed a prediction of paleontology: not only was the new fish an intermediate between two different

> kinds of animal, but we had found it also *in the right time period in earth's history* and *in the right environment.* (Shubin 2008, 23–24, emphasis in original)

If modern evolutionary theory and geological dating methods are the farrago of nonsense the creationists claim, then we must conclude that Shubin's team was incredibly lucky.

Every year thousands of papers are published in professional journals of evolutionary biology and its related fields. Each one of those papers represents a small dent in the sum total of human ignorance and is a success for the evolutionary approach to natural history.

For me, this is the central fact of the matter. Armchair argument can be stimulating, but it is ultimately beside the point. Scientists are expected to produce results, and toward that end they are forced to be pragmatic. They rely on what works. Evolution survives because it routinely passes the only test that matters to anyone who has ever entered a laboratory with the hope of discovering something new. For more than a century it has produced tangible progress. Creationism offers nothing to rival it.

11

Fossils, Human and Otherwise

To this point science had played only a small role in the proceedings of the Creation Mega-Conference. For me, that changed with the next talk that I attended, "Rocks Around the Clocks: The Eons That Never Were." The speaker was Emil Silvestru, a geologist originally from Romania. (This talk was in the advanced track. His competition, in the basic track, was another round with Ken Ham, "How to Defend the Christian Faith in a Secular World." Based on long experience, my strong suspicion was that the proper mode of defense would involve an uncompromising approach to Genesis.)

Silvestru's argument was that such geological evidence as we have entirely supports the young-Earth hypothesis. He said:

> If creation is true you would expect that there are no intermediate forms ever. There's no transitional fossils, all have been created according to their own kind. And you would expect the vast majority of them to be buried in water-laid sediments.
>
> If evolution is true you should find a lot of transitional forms, transitional fossils, and some of them buried in sea sediments, some of them buried in continental or dry-land sediments.
>
> Now you go to the facts and you notice that 99 percent of all existing fossils are sea creatures killed or covered by sea sediments, and there's no transitional forms.[1]

The blunt assertion that there are no transitional forms, common in the literature of creationism, is the sort of thing that justifies the contempt of anti-creationists. A quick browse through a textbook on paleontology should disabuse you of any positive opinions regarding the claim's credibility. We saw a striking example of such a fossil in the previous chapter, linking fish to land-dwelling animals. There are also transitionals bridging fish to amphibians and amphibians to reptiles. The fossils linking reptiles to mammals are so numerous and well-preserved that it is common to speak of mammal-like reptiles.[2] The collection of fossils documenting the appearance of humans from ape-like ancestors is large and continues to grow.

The evolution of whales from land mammals is documented by an impressively detailed sequence of transitions. Those are just a few items. There are many others.[3]

Regarding Silvestru's other claim, there is a simple explanation for why most fossils are found in water. Fossilization can only occur under special circumstances. A dead animal quickly becomes food for scavengers, and the remains do not last long in the wild. Fossilization requires rapid burial by sediments, which is far more likely to happen in the sea. Biologist Kenneth Kardong writes: "Animals living in water or near the shore are more likely to be covered by mud or sand when they die than are upland animals. Upland animals die on ground exposed to scavengers and decay. Thus, most fossil-bearing rocks are sedimentary—formed in water" (Kardong 2005, 30).

Thus, in Silvestru's statement, we see two standard rhetorical techniques. In the first, false claims are put forward bluntly and with confidence. In the second, entirely mundane observations are presented as though they posed a difficulty for conventional theory.

This, mind you, was only from the presentation's opening. From here, Silvestru proceeded in bullet-point fashion, tossing off one example after another meant to persuade us of the bankruptcy of modern geology. I fear it would strain the reader's patience beyond decency to discuss each of his points in turn. Since they were comparable in merit to what we have seen thus far, we shall move on.

The next talk, "Ape Men, 'Missing Links' and the Bible," by Phillip Bell, a speaker from England with a background in cancer research, featured nuggets like this:

> In an evolutionary view, if we're logical, if we're consistent, then it leads to a belief that there is no purpose for your existence. And that leads many people down the path of nihilism, fatalism, believing that if things get tough then suicide's an option. I think it's no surprise that suicide is the leading or second leading cause of death in many young people today in Western countries. We can set our own rules, many people say. And their justification is, we're not accountable to any God, any creator, we're just evolved animals.
>
> Let me say right at the start, for Christians, this is not an optional issue. It's one of the key things which is at the heart of what makes us human and whether we are made in the image of God or not.[4]

With this melodramatic statement the conference finally got down to business. It is really human evolution that makes creationists bristle with disgust.

I happen to agree that there is tension between evolution on the one hand and traditional notions of divine creation on the other. More difficult to fathom, however, is the leap from evolution to nihilism and suicide. Why should my attitudes towards the value of life or my own self-worth be governed by the manner in which I appeared? The idea that life has no meaning if God does not exist is something creationists believe with passion and sincerity, but it is an attitude I fear I shall never comprehend.

Metaphysical theorizing notwithstanding, the fossil evidence for the evolution of humanity is really quite strong. Biologist Jerry Coyne provides a useful summary:

> Around five to seven million years ago, then, we expect to find fossil ancestors having traits shared by chimpanzees, orangutans, and gorillas (these traits are shared because they were present in the common ancestor), but with some human features too. As the fossils become more and more recent, we should see brains getting relatively larger, canine teeth becoming smaller, the tooth row becoming less rectangular and more curved, and the posture becoming more erect. And this is exactly what we see. (Coyne 2009b, 196)

Bell offered two main arguments for rejecting the fossil evidence of human evolution. The first was that all alleged intermediates are either fully human or fully ape. Thus, the Australopithecines, among the oldest hominid species known from the fossil record, were dismissed as merely a variety of ape. More recent fossils, such as *Homo erectus*, were classed as human beings. The argument was made that in some cases the fossils we have represent the victims of bone diseases, which accounts for their somewhat corrupted appearance relative to modern humans.

Seen in this light, claims regarding the wealth of ape-human transitional forms are based on the exaggerations and evolutionary blinders of secular paleontology. Consider, for example, Bell's comment on an exceptionally complete fossil skeleton known as "Lucy." (As he was saying the following, he displayed a slide with Lucy on the left and a modern human skeleton on the right.)

> Since more material has been found they've found out that this creature was fully ape-like. Ape-like stature, three to four feet tall, grasping hands and grasping toes, would have walked on its knuckles when he was walking on all fours, probably climbed trees quite adequately, and so on. You can see on the right-hand side, the reconstruction, a more conical rib cage much more like apes today. Nothing like the human rib cage. So Lucy is clearly an ape based on the actual material we've uncovered.

You will not find many scientists agreeing with this assessment, but not because of any bias on their part. The problem is simply that Bell's casual assertions are not supported by the facts. In a recent study, biologist Phil Senter studied thirty-six anatomical characteristics of the Lucy fossil (known as AL 288-1 in the professional literature). He summarized his findings as follows:

> Of 36 anatomical characters examined on AL 288-1, 14 (39%) exhibit the apelike state and 22 (61%) exhibit the human-like state. Of the 22 characters for which AL 288-1 exhibits the human-like state, 12 (55%) are found on the vertebral column,

> pelvis, and lower limbs and could therefore be construed as related to bipedal locomotion, whereas 10 (45%) are found on the jaw, teeth, and upper limbs and are therefore unrelated to upright bipedal locomotion. The data therefore support the hypothesis that Lucy is anatomically intermediate between apes and humans and falsify the hypothesis that Lucy is just an ape. (Senter 2010, 71–72)

We must also remark on the colossal disrespect shown to scientists by arguments such as Bell's. Peruse the professional literature on paleontology, and you are struck immediately by the care and precision that goes into this sort of work. Conclusions that fossil hominids show a mixing of human and ape characteristics, appearing in a sequence consistent with evolutionary expectations, are not the result of cursory examinations or naturalistic biases. They are instead the consistent findings of hundreds of professionals.

Bell's second argument for rejecting the fossil evidence for human evolution involved disagreements among paleontologists regarding the proper classification of hominid fossils. As we have noted, fossils are just snapshots. From them we learn there were animals possessing certain traits at a particular time in natural history. Each fossil is the remains of a creature that, when it was alive, was a representative of a population of hominids. Was *that* the population that gradually evolved into modern humans? Or was it instead a mere side branch?

Such questions are impossible to answer with certainty. We can, however, make detailed comparisons of the traits preserved in the fossils to make determinations about relative relatedness. That is, we can make judgments that species A and B are more closely related to each other than either is to species C. This is difficult, painstaking work, and all such judgments must be considered tentative. Periodically new fossils are discovered to shed light on questions where previous opinion had been stagnant.

Bell thought all of this was hugely significant. He relentlessly hammered the point that scientists themselves do not always agree on the proper classifications. It is not difficult to compile quotations from scientists challenging, sometimes vehemently, the interpretations of other scholars. Bell had many such quotations to offer.

This is a weak argument, though, because it confuses two separate questions. One is whether we can be confident that evolution has occurred. The existence of numerous transitional forms connecting humans to ape-like ancestors is strong evidence that it has. A different problem is reconstructing the actual course of natural history. That is a far more difficult question, and it is one whose answer is constantly revised as new data becomes available. Statements acknowledging the difficulty of this problem are not relevant to the soundness of evolution generally.[5]

One of the least endearing features of creationist discourse is the sheer magnitude of the charges they direct towards evolutionists. It is insufficient for their purposes merely to persuade audiences that evolution is factually mistaken. They also feel the need to link evolution to every type of nastiness ever to afflict humanity. We saw an example of this at the start of Bell's talk, where evolution was causally linked to nihilism and suicide, and we may as well close in the same way. Bell was discussing fossils discovered in 2003 that represent a hominid species known as *Homo floresiensis*. There has been speculation that these fossilized creatures—dubbed "hobbits" in the media on account of their diminutive stature—represent a sidetrack in human evolution. This led Bell to comment:

> If they were descended from *Homo erectus*, and *Homo erectus* was not human, then do you know what? . . . That would mean they're not descendants of Adam. And what would that mean? They can't be saved. There'd be no point sending missionaries to them would there? Because they can't be saved unless they're descendants of Adam. If you read 1 Corinthians 15 and Romans 5, the apostle Paul makes that clear. Evolution is a racist idea, you see. You need to be so careful.

A remarkable monologue. I pass it along without further comment.

12

On Information

On the second day of the Creation Mega-Conference, I attended a math talk, "In the Beginning Was Information," by Werner Gitt. The drama that occurred after the talk made the conference personally memorable for me, but we shall come to that in due course.

In everyday life the word "information" connotes the idea of a meaningful message. You ask your colleague in the neighboring office what time the meeting starts. He replies, "Two o'clock." Information has been transmitted from a sender, your colleague, to a receiver, yourself. Were I to show you a page covered in hieroglyphics you might say, "These symbols mean nothing to me." The symbols on the page do not become information until there is a suitable receiver to give them meaning.

Information might be transmitted through some physical medium, such as patterns of ink molecules on a page or vibrations in the air caused by someone's voice, but the information itself is nonmaterial. You can understand perfectly the physical principles by which ink adheres to paper without apprehending the information the ink conveys. Meaningless scribbles and meaningful words are governed by the same physics, but there are critical differences between them nevertheless. Moreover, information is independent of its substrate. The same information can be carved in stone, written in ink on paper, shouted across a room, or stored on a hard drive.

That information is immaterial warms the hearts of creationists. As they see it, information is a fundamental attribute of living systems, and, since it cannot be understood in purely material terms, it forces us to introduce a spiritual dimension.

Biologists speak casually of "the genetic code," but a code implies a representation of information. Human genes contain the information for building a human being, horse genes for building horses, and so on. According to evolution, genetic information has increased through the course of natural history. If natural processes cannot adequately account for this growth, then evolution, at least as conceived by scientists, must be discarded.

It was Gitt's project to transform these vague and imprecise musings into a bona fide argument. The conference program included detailed lecture notes for his presentation.[1] Gitt summarized his goals:

> We will set out in a new direction, by seeking a definition with which it is possible to formulate laws of nature about it. Information is a nonmaterial entity and this is the first time that a law of nature has been formulated for a mental concept. First, we will describe the distinguishing attributes of information, formulate its definition, state the laws themselves and draw six strong conclusions. Since we have successfully discovered and formulated ten laws of nature about information, we will refer to this definition of information as Laws of Nature about Information (LNI).

Such grandiose claims ought to trigger whatever skeptical impulses you possess. A person claiming to have discovered ten new laws of nature might be an unheralded genius, but he is far more likely to be kidding himself. When those laws are said to be based on novel concepts that have eluded professionals, the probability of the latter explanation rises.

I had gotten that far upon arriving at Gitt's six strong conclusions. Here they are, verbatim:

- God exists; refutation of atheism.
- There is only one God, who is all-knowing and eternal.
- God is immensely powerful.
- God is spirit.
- No human being without a soul: refutation of materialism.
- No evolution.

I fear we have now reached the point where we can dismiss things out of hand. There is no way that an abstract, mathematical model could possibly, even in principle, yield conclusions *that* momentous.

Still, since "information" plays so central a role in the discourse of both creationism and intelligent design, it behooves us to consider things a bit further.

There is a branch of mathematics known as "information theory." It was pioneered in 1948 by Claude Shannon, then a researcher at Bell Labs. That information can be given a mathematical treatment might seem paradoxical in light of our previous discussion. Mathematics requires precise, objective definitions, but information in the everyday sense is inescapably subjective.

Shannon noticed that something quantifiable remains after you have removed the aspect of meaning from our everyday notion. We begin by imagining that one of several possible messages will be transmitted across some communication channel. Each message has a certain probability of being transmitted. We are initially uncertain which message will be transmitted, but after receiving the message, that uncertainty is gone. It is this reduction in uncertainty that is precisely quantified within Shannon's theory. The idea is that a state of maximal uncertainty is represented by a situation in which all of the possible messages have the same probability. By contrast, when one message has a vastly greater probability of occurring than any of the others, there is far less uncertainty regarding which one we shall receive.

The motivation for this came from certain practical problems regarding the efficient transmission and storage of information. Like all really good ideas, however, Shannon's work quickly found applications in other areas. One such is biology, and nowadays "bioinformatics" is a well-established branch of science. Researchers in this area rely heavily on mathematics and computer science, including Shannon's information theory. An animal's genes can be viewed as a set of coded instructions. It therefore seems meaningful to say they contain information.

This can lead to confusion, since the word "information" is used by biologists in two different ways. Formally it refers to the application of Shannon's ideas to specific biological questions. Informally it is just a metaphor expressing the notion that genes convey a meaningful message. Philosopher Paul Griffiths writes:

> It is conventional wisdom that insofar as the traits of an organism are subject to biological explanation, those traits express information coded in the organism's genes. . . . I will argue, however, that the only truth reflected in the conventional view is that there is a genetic code by which the sequence of DNA bases in the coding regions of a gene corresponds to the sequence of amino acids in the primary structure of one or more proteins. The rest of "information talk" in biology, and the claim that biology "is, itself, an information technology," is on a par with the claim that the planets compute their orbits around the sun or that the economy computes an efficient distribution of goods and resources. It is a way to talk about correlation that, in some cases, allows a useful application of the mathematical theory of communication and in others plays no theoretical role but merely reflects the current cultural prominence of information technology. (Griffiths 2001, 395)

Thus, information can be spoken of in a formal, mathematical way, but only by sacrificing the aspect of meaning central to the everyday use of the term. Alternatively, you can speak of information casually to refer to a meaningful message, but then you must abandon any hope of a precise, scientific discussion.

Or so it seems to most researchers. Gitt demurred. Having no use for the distinctions I just discussed, he provided a definition of information that included subjective notions like "meaning" and "purpose." He then asserted that where information is found it must be the result of an intelligent agent. Near the end of his talk, he said,

> With the help of the laws of information we can reach a comprehensive and fundamental conclusion: the idea of macroevolution—i.e. the journey from chemicals to primordial cell to man—is false. Information is a fundamental and absolutely necessary factor for all living things. But all information—and living systems are not excluded—needs a nonmaterial source. The evolutionary system, in the light of the laws of information, shows itself to be an "intellectual perpetual motion machine."

When Gitt concluded his talk, the auditorium erupted into enthusiastic applause. Before long, everyone except me was standing. Ken Ham took the stage and boasted that this was one of the most powerful apologetic arguments he had ever heard.

It was a reminder of something I mentioned in chapter 2. Most of my fellow audience members had little to no training in math or science. Consequently, the talk they heard was very different from the one I heard. Whereas I heard a caricature of a well-established branch of mathematics, they heard welcome scientific confirmation of their religious beliefs.

The trouble was that Gitt's argument, for all its technicalities and liberal use of jargon, ran afoul of a few simple facts. When scientists attribute to the evolutionary process the ability to increase genetic information, they are not speculating or guessing. They can point to known, well-understood mechanisms that demonstrably have that power.

One such process, described in detail in most genetics textbooks, is gene duplication and divergence. The principle is straightforward. In high school biology you learned about the mechanisms through which genes replicate themselves. Sometimes these mechanisms lead to the duplication of a gene. As a result, there are now two copies of the same gene. The second copy is free to accumulate mutations while the original can continue performing its usual function. The combination of duplication with subsequent divergence plainly leads to the creation of new information.[2]

Studies of natural selection both in the field and the lab have established its ability to adapt organisms to their environments. It is an observational fact that populations can evolve novel functionalities by passing random genetic variations through the sieve of natural selection.

None of this is mysterious, hypothetical, or controversial. Nor is it cutting-edge. An understanding of the importance of gene duplication in evolution dates back to at least 1911 (Taylor and Raes 2004). The study of natural selection specifically in information-theoretic terms also has a long history (Kimura 1961). If your abstract, mathematical model says that genetic information cannot increase through natural means, then your model is simply mistaken.

As with many of the previous talks, there was no formal question period. However, a large crowd formed around Gitt to speak to him. I joined them and worked my way towards the front. I would estimate the crowd to be around forty people.

Unsurprisingly, most of the questions were supportive of Gitt. At one point he started gushing about how all it would take to refute his ideas was for a scientist to produce a single natural mechanism that could increase the information content of the genome. I decided it was time to break cover. With a bit more volume and heat than intended, I said, "How does a single mutation change the information content of the genome?"

Silence—as forty pairs of eyes turned towards me. I swallowed hard, forced myself to stay calm, and continued: "As I'm sure you're aware, genes mutate all the time. Before I can answer your challenge, I need to have a better understanding of your notion of information. So tell me how a simple point mutation changes the information content of the gene."

One of the mainstays of creationist argumentation is that all mutations lead to "a loss of information." I use the sneer quotes because they almost never formulate the concept with sufficient precision to know what they mean. It seems obvious, for example, that the result of a typical point mutation (by which we mean a single change to one DNA base) is that the gene will no longer produce the protein it previously made, and will instead produce a different protein. Does that represent a gain or a loss of information?

Sure enough, Gitt told me that mutations always degrade information. In the heat of the moment, I did not think of the obvious counter. It is a logical impossibility for all mutations to degrade information. Mutations can reverse themselves, you see. If the mutation changing A into B causes a loss of information, then the reverse mutation from B back to A must represent a gain of information.

Instead I said: "You keep talking about information going up or information going down. You talked about the cell containing more information than an encyclopedia. But at no point did you tell us how to measure information. And without such a measure it's not even meaningful to talk about information content increasing or decreasing." I then added, "Usually when scientists talk about information they have in mind Shannon's concept. When it comes time to measure information, is that what you have in mind?" Gitt had briefly discussed Shannon in his talk.

There was a high school student standing near Gitt. He was not amused. He told me, a bit rudely I thought, that I should consult *The Answers Book*, by which he meant a specific tome on sale in the bookstore. He assured me it would answer my question.

It is helpful, when attending these conferences, to be very well-read in creationist apologetics. I had read that particular book and knew it did not address my point. So I flashed him my most withering look and said: "I've read that book and it does not answer my question. Now if you don't mind I'd like to hear what Dr. Gitt has to say."

Gitt hemmed and hawed but eventually conceded that information can only be quantified at the "statistical" level and that for the purposes of measuring information that is what was important.

So I replied: "If that is what you mean, then there are several well-known mechanisms that can lead to an increase in information content. Here's one." I explained the process of duplication and divergence. I asked, "Why is this not an adequate response to your challenge?"

His reply was that duplicating a gene does not create new information. It is like photocopying a page out of a book. It leaves you with more of the same, not anything new. My attempts to make him realize that the process involved not only duplication but also divergence fell on deaf ears.[3] The conversation was becoming a bit chaotic, since many people were now attempting to put me in my place. It was a standard exercise in futility, but at least I made it clear that there are answers to his arguments. There was still a large crowd waiting to ask questions, and I did want to monopolize his time.

I shook his hand, thanked him for his time, and started to walk away. Mentally I was already thinking about lunch, for which I had picked out a small Mexican restaurant near the university. I was quickly calming down from the excitement of the moment and was ready to move on to other things.

That was when I heard someone behind me say: "You're really very ignorant about biology. You should learn a bit more before you start talking about it." Lunch, it seemed, would have to wait. My new tormentor was a woman in her late thirties. She was there with her son, who looked about twelve.

When I first starting teaching mathematics, just over a decade ago, I received a piece of advice from a senior faculty member. He told me that no matter how clear you think you are, no matter how impossible you think it is that anyone could misunderstand you, there will be times when you are asked a question that leaves you stunned. It is easy to forget just how strange mathematics looks to people who are not immersed in it.

I was reminded of that during the ensuing conversation. There was simply no communication going on between us. To someone watching from afar it would have seemed we were involved in a discussion, but that would have been a serious misapprehension. I would speak for a while, and she would nod at the appropriate moments, but then she would come back with something incoherent and irrelevant to what I had just said. I recognized in her remarks a number of standard talking points, but they were garbled to the point of being barely recognizable.

At no point, however, did her confidence flag in the slightest. She seemed sincere in her belief that she was entirely on top of things. She was bolstered in that perception by the half dozen or so other people who gravitated over to join us. They also had firm ideas regarding who was representing light and truth at that particular moment. I spent as much time helping them state the creationist arguments clearly as I did explaining why those arguments were mistaken.

I am sorry if this seems arrogant or smug, but there really does come a point where you must insist that facts are facts. We were not arguing about subtle issues of evidential interpretation or nuanced philosophical questions about worldviews or biases. They were simply using jargon whose meaning they did not understand, and tossing off assertions that bore little resemblance to the findings of modern science. They were utterly contemptuous of scientists and refused even to consider that the story was more complicated than their simplistic catchphrases allowed. Whatever arrogance or smugness there was in the conversation was entirely on their side.

On the other hand, let's not be *too* grim. You see, there was one exception to the general irritation I felt toward my sparring partners. I am referring to the twelve-year-old son of the woman who taunted me. He seemed genuinely interested in what I was saying, and when he interjected a comment it was always something intelligent. As I shook hands with the people around me and finally left to round up some lunch, I amused myself by thinking of the serious teenage rebellion he was soon going to inflict on his mother.

13

Movies and Television

Later that afternoon, the basic track featured another presentation by Carl Kerby, this time called "Evolution and Pop Culture."[1] The talk was primarily a series of clips from movies and television that in some way mentioned evolution or the great age of the Earth. We were encouraged to be ever vigilant for such things, and to use them as teachable moments with our children. For example, with regard to the 2002 film *My Big Fat Greek Wedding*, Kerby writes,

> My daughter called me from college and said, "Dad, you're not going to believe it. There's evolution in *My Big Fat Greek Wedding*!" This was very encouraging to us as parents. We had spent time with our children training them to catch these subtle influences, sometimes wondering if they were having any impact. This was confirmation that not only had our children been listening, even though they may have acted as if they weren't, but they were also learning.
>
> Anyway, where's the evolution? Well, it happens when Toula's (the main character's) family is celebrating Easter and her fiancé tries to say "Happy Easter" in Greek to her dad. Her dad is not impressed and says back to him in Greek, which is shown in subtitles, "When my people were writing philosophy, your people were still swinging from trees." Now that's not very subtle, is it? (Kerby 2006, 36–37)

Such things are annoying in films directed toward adults, but they become insidious when directed toward children. A case in point is the 1940 Disney film *Fantasia*, which includes a lengthy depiction of evolution. Kerby writes: "You see, this goes all the way back to 1940. Evolution has been indoctrinating through film, and destroying our self-worth at the same time, for a long time" (ibid., 50).

In some cases the attempted indoctrination was subtle to the point of being subliminal. Such was the case in the 2004 animated film *The Incredibles*. In one scene a character is reading a newspaper. If you look closely you can see that one of the headlines says, "Scientists Find Missing Link" (ibid., 55).

Kerby had many other examples as well, and the audience was by turns amused and aghast.

Walking out of the room, I was inclined to scoff at Kerby's examples. To me he seemed awfully worked up over a lot of trivia. But the more I thought about it, the more I came to think he had a point. I thought about all the times I had been annoyed by gratuitous Christian sentiment in movies or television shows that were not overtly religious in content. No doubt more religiously inclined people would think I was the one who was overreacting. In the end the talk served to remind me again of just how isolated creationists feel in pondering popular culture.

As it happens, that was not the only time I heard about Darwin's pernicious effect on the entertainment industry. At an earlier Intelligent Design conference, held near Kansas City in 2002, I was in the audience for a presentation called "Darwin Does Hollywood: Naturalism in the American Movies." The speaker was named Jack Cashill.

Whereas Kerby was worked up about overt references to evolution or the antiquity of the earth, Cashill's concerns were more subtle. The bulk of his presentation was taken up with a series of film clips, most of them intended to horrify us with their casual immorality and naturalistic philosophy. We learned that in its early days, the film industry, worried about the trend toward cinematic debauchery in the 1920s, devised the Hays Code to keep the slide in check. The code's censorship requirements were respected by nearly all American studios. Among its line items were prohibitions against ridiculing religion or religious figures and requirements that evil characters not be presented sympathetically. The code was in place from 1930 through 1968.

Cashill opened with a clip from the 1948 film *Rope*, directed by Alfred Hitchcock. The plot involves two arrogant college students who murder a classmate they regard as intellectually inferior. They have no motive for the crime beyond their own belief they can get away with it. Eventually the students are found out by their former professor. The teacher is horrified to learn that some of his overly casual musings about social caste and intelligence had given his students the idea for the crime.[2]

In praising the film Cashill said,

> This movie was made while the Hays Code was still in effect, so the Darwinian characters are bad. They're also obviously homosexual; they're played by homosexual actors, too. Homosexual critics criticized the movie because it was one of the rare times in Hollywood where homosexuals are seen in a negative light. I think it's one of the few times Darwinians are seen in a negative light.

Sadly, with the demise of the Hays Code in 1968, films were free to return to their decadent roots. An example was the 1972 film *High Plains Drifter*. In this Western, starring Clint Eastwood, an unnamed drifter wanders into the small town of Lago. He soon becomes involved in a gunfight, with questionable provocation, and kills three people. Impressed by his skill, the town's leaders hire him for protection.

Three violent convicts have just been released from prison, and we learn that these men blame the town's people for their incarceration. Eastwood is offered anything he wants in return for his services. Though he repeatedly humiliates the residents, their fear of the approaching danger outweighs the damage to their pride.

Cashill's objections to this film were numerous, including the frequently foul language and favorable presentation of the drifter's less-than-admirable behavior. Also problematic is the presentation of the town's preacher:

> The preacher character in *High Plains Drifter* is, of course, like all preacher/religious characters in the last thirty years, a buffoon. And he walks into this bar, and he tries to intervene in this bloodletting, and they tell him—and this speaks for Hollywood, too—"Take your conscience elsewhere." And Hollywood took its conscience elsewhere.

Cashill compared *Drifter* unfavorably with the film *High Noon*:

> Compare this movie in your mind to *High Noon*, which was made twenty years earlier . . . same basic plot, three bad guys escape from prison, one man has to stand up against them. Now if you have the image of Gary Cooper in your head, totally moral, married to a Quaker, defending the town, going to the church. Now compare him to the character Clint Eastwood plays here.

Cashill's presentation was marked throughout by casual equivalences drawn among communism, naturalism, greed, immorality, atheism, degradation of people, and poor treatment of women. All of this was linked in some vague way to the Pandora's box opened by Darwin when he banished God from natural history. A classic example is the following statement about Martin Ritt, a prominent film director whose career spanned thirty years, from the 1950s to the 1980s. Cashill was referring to Ritt's 1963 film *Hud* :

> It's also, in fact, Martin Ritt, the director, who was blacklisted as a Communist. Communists didn't make communist movies in America, they merely made movies without God in them, which was their kind of a compromise, which is why Marx found Darwin so attractive.

To go so quickly from communism to Darwin—through atheism and Marx no less—is no small feat. I invite you to imagine the effect such concentrated use of buzzwords had on the sympathetic audience.

I am always impressed by the sheer viciousness of so much anti-evolution rhetoric.

Cashill's remarks about *Rope* are a case in point. His casual description of the film's murdering, amoral characters as "Darwinian" leaves one speechless. Any thoughtful person, in pondering the relationship between accepting evolution and finding it agreeable to commit murder based upon personal feelings of superiority, would have to conclude that no connection exists. You must constantly remind yourself that creationists arrogate to themselves the right to use language in any manner they choose. "Darwinian," to them, is just a synonym for atheism, which in turn is synonymous with immorality and a craving for sin. That everyone outside their subculture thinks the term has something to do with biology reflects a flaw in *their* understanding.

Notice also Cashill's comments about homosexuals. The sole connection between homosexuality and Darwinism is, so far as I can tell, that conservative Christians dislike them both. Also, it is sheer fantasy to suggest that negative portrayals of homosexuals in film have been historically rare. The reality is that favorable portrayals are a very recent phenomenon. Prior to the last twenty years or so, it was all but universal to portray homosexuals as sissies or as morally untrustworthy (Russo 1981).

Similarly blinkered is the notion that films made under the Hays Code were necessarily paragons of moral virtue. The code prohibited depictions of interracial marriage, for example, and films of the time routinely portrayed women and minorities as second-class citizens. "Golden Age" myths, in which a degraded present is contrasted unfavorably with a glorious past, are usually gross oversimplifications. That is certainly the case here.

I will need to beg your indulgence now, since I cannot simply ignore such casual denigration of a cinematic masterpiece like *High Plains Drifter*. Far from being a celebration of immorality or being symbolic of Hollywood's having taken its conscience elsewhere, *Drifter* can far more plausibly be cast as a Christian parable. It is a story about the wages of sin and our inability to run indefinitely from our iniquities.

I should probably warn you there are spoilers ahead.

Comparisons to *High Noon* are appropriate, but not for the reasons given in the presentation. *High Noon* is the story of a lone sheriff, played by Gary Cooper, courageously doing his duty in defiance of both the overwhelming odds against success and the cravenness of the people he is protecting. Cooper is the focus of the story. The moral vacuum of the town exists mostly as a device for spotlighting Cooper's greatness.

Drifter is what *High Noon* would have been had it been told from the perspective of the town. It is Lago itself that is central to *Drifter*'s story, not Eastwood's character. The point is not to cheer the drifter's generally poor behavior. It is to marvel that the town folk are so wretched, so trapped by a desire to conceal their past crimes, that they are utterly unable to stand up to the humiliation Eastwood heaps upon them.

As the story opens we seem to have a simple tale of a decent town menaced by an outside threat. As things progress, however, we learn the approaching criminals have a legitimate complaint. The town's sheriff had uncovered criminal activity in the local mining company. The company responded by hiring the three thugs to murder the sheriff. The murder took place in the center of town, with most of the residents watching. No one lifted a finger to help the sheriff, since they knew the mining company was the town's economic lifeblood.

Armed with that background, let us revisit the scene in which the preacher was told to take his conscience elsewhere. It did not take place in a bar, but during a meeting of the town's most prominent citizens. The preacher protests that his conscience will not permit him to hire a killer, and he describes himself as "a simple man of God." The mining company's leader replies,

> Well it's time we unsimplified you Reverend. [The three men the drifter killed] were professional gunfighters on the payroll of the Lago Mining Company to protect our interests, and the interests of the town, which are identical. They stood around drinking beer and looking snotty for a full year. And then, one day before we actually needed the bastards, they went and got themselves killed. So if you've got a suggestion we'd be delighted to hear it. Otherwise, take your conscience elsewhere while we think about saving your ass.

The preacher quickly backs down.

No one watching that scene could think the preacher's tormentors were acting honorably or were being presented nobly. The scene appears early in the film, prior to the later revelations about the secrets the town is hiding. It is intended to make the audience realize that the notion of decent town folk menaced by external threats is far from the whole story.

This sort of gross oversimplification was typical of the presentation generally. In every case the films under consideration were far richer and more nuanced than could be captured in a few out-of-context scenes. Creationists, alas, are not fond of nuance. They prefer stories that are stark and simple.

PART III

The “Darwin vs. Design” Conference, Knoxville, Tennessee, March 2007

14

The Marginality of Genesis 1

To judge from my experiences in Lynchburg, you might think that anti-evolutionism is largely about Bible-thumping obscurantism. This is a serious misapprehension, as demonstrated by my experiences at a conference called "Darwin vs. Design," held in Knoxville, Tennessee, in March 2007. This conference focused on intelligent design, a relatively new strain of anti-evolutionism that became prominent in the 1990s. While the beliefs of my fellow attendees certainly had much in common with those I encountered in Lynchburg, there were pronounced differences as well.

Over the years I have asked a great many creationists the blunt question, "What do you find so objectionable about evolution?" Never once has anyone replied, "It contradicts Genesis." Biblical questions often arise later in the conversation, but they are never the paramount concern. The major problem is really the effect of evolutionary theory on perceptions of human significance. Having accepted that we are the result of a naturalistic and unpredictable evolutionary process, it is difficult to accept simultaneously that we are the intended creations of a loving God. From this general concern follows numerous, secondary issues. How are we to understand the soul in the light of evolution? Is a process as cruel and wasteful as Darwinian evolution really the sort of thing a loving God sets in motion? What foundation do we have for morality in Darwin's world?

Some Christians have resolved these questions to their own satisfaction. Many others have not, however, and the polling data we saw in chapter 7 suggests that a majority of *them* are not YECs. They have religious objections to evolution, but not because of dubious attitudes toward geology or idiosyncratic interpretations of the Bible. This attitude has a long history among Christian thinkers, going back to Darwin. A consideration of that history will help us understand the beliefs of modern proponents of intelligent design.

In 1802, William Paley, a British philosopher and clergyman, published *Natural Theology, or Evidence of the Existence and Attributes of the Deity, Collected from the Appearances of Nature*. His central argument was famously expressed in the book's opening paragraph, in which we are asked to contrast a stone with a pocket watch. Asked about the origin of the stone, we could shrug our shoulders and say that for all we know it might have lain there forever. We would not so respond regarding the watch. The clear time-telling purpose achieved by the intricate arrangement of gears and springs clearly calls for a special kind of explanation. Specifically, for intelligent design. Paley elaborated on this theme by providing an erudite tour through some of nature's more impressive productions.

As Paley himself acknowledged, there was little new in the substance of his book. The design argument had a pedigree going back to antiquity and had been well-explored in other works. Paley provided extra detail and novel examples, but the basic argument was familiar.[1] You might wonder, then, why he wrote the book at all.

The reason is that Paley's motives were only partly apologetic. Paley was living in the wake of the French Revolution, commonly viewed as an expression of an atheistic political philosophy. Heretical writing was common among philosophers of the time, and evolutionary speculations were also not unheard of. That Paley wished to counter such trends is hardly in doubt, but he had other motivations as well (O'Flaherty 2010).

The detailed study of nature was, for him, a form of religious devotion. Such contemplations led you intellectually to the fact of God's existence, but emotionally they brought you closer to Him. This was especially important for a younger generation in danger of losing its faith to the skepticism of the day. Historian Niall O'Flaherty writes:

> Though few eighteenth-century thinkers questioned the validity and importance of natural theology, there was disagreement among churchmen as to its proper function. . . . What was new about Paley's book, as we shall see, was that it integrated an advanced inculcatory psychology into natural theology in an attempt to further unlock this devotional potential. The key point is not simply that *Natural Theology* had a devotional function, as has frequently been observed, but that for Paley it represented piety in its purest form. Most importantly—and this is why he was so keen to make them a more prominent part of every Englishman's daily devotions—Paley recognised that theological nature studies, if presented in a manner that gave due regard to the workings of the mind, might give rise to a substantially more virtuous disposition. (ibid., 21)

The devotional aspect and emotional appeal of Paley's work is often overlooked in anti-creationist literature. Among those proposing reconciliations of evolution with Christianity, it is common to downplay the significance of the design argument's demise after Darwin. No central Christian tenet rides on the correctness of Paley's reasoning, it is argued. Moreover, we have the option of shifting the argument back one step. We might no longer find design in the contrivances of organisms, but we can find it instead in the exquisite system of natural laws that makes possible the existence of life in the first place.[2] At any rate, belief in God is about faith. While it is reasonable to see reflections of divine glory in nature, perhaps we should not be thinking in terms of gathering evidence for God's existence.

These are cogent points as far as they go, but I can understand why so many Christians find them unsatisfying. The design argument is only partly about proving God's existence. Equally important is its capacity for emphasizing the nearness of God in our everyday lives. It is a line of thought available to anyone, for you hardly need a degree in science to appreciate that animals are complex and that design needs a designer. Paley understood this, which is why he viewed the serious contemplation of such things as a device for strengthening faith.

Now here come the Darwinians to say that this most fundamental spiritual instinct is flatly mistaken. The design we perceive in organisms is solely the result of natural selection, they say. If it represents the culmination of a divine plan, then it is nevertheless one that was set in motion billions of years ago. Transfer the design argument to the recondite land of modern cosmology, and it loses all of its emotional power. The complexity of organisms is obvious to all. In this it differs from the minutiae of the cosmological constant or the vagaries of quantum mechanics, which are things we only hear about in physics courses.[3] As viewed by many modern anti-evolutionists, something more than a mere intellectual argument has been lost by discarding Paley's reasoning.[4]

Try to imagine how it looks to them to read statements like this, from theologian Peter Hess: "What are the central theological failings of intelligent design? First, it is blasphemous. Intelligent design constrains God to work within the limits of what its adherents can understand about nature. In so doing, it reduces God from the status of creator to that of mere designer . . ." (Hess 2010, 104). Intelligent design proponents, of course, would not agree that they are limiting, constraining, or reducing God. But they would be downright amazed by the suggestion that, in suspecting that certain aspects of the natural world are best explained by reference to God's actions, they are guilty of blasphemy. As Paley showed us, seeing reflections of God's designing hand in nature is a way of drawing closer to Him. It is hard to believe God would be insulted by such efforts.

In *On The Origin of Species*, published in 1859, Charles Darwin provided a plausible naturalistic explanation for biological complexity. It is common to recognize two distinct purposes in Darwin's work. The first was to convince people of common descent. The second was to establish natural selection as the primary mechanism. Among scientists he was far more successful with the first purpose than with the second. Common descent quickly won the day following the book's publication. Natural selection, by contrast, would have to wait until the 1940s for its eventual triumph.[5]

To a surprising degree, the religious response during this time paralleled the scientific response. Natural selection was a far more bitter pill to swallow than was common descent. There were certainly some among the faithful who wanted nothing to do with any of Darwin's ideas, and the popularity of such attitudes should not be minimized. But, especially among intellectuals, it was generally natural selection to which people had specifically religious objections.

Representative are the views of theologian Charles Hodge. From 1851 through 1878, he was the president of Princeton Theological Seminary. The influence he held over Protestant thought, both from his professional position and through his voluminous writings, grant a special importance to his views. His final book, published in 1874, bore the title *What Is Darwinism?* His blunt answer:

> We have arrived at the answer to our question, What is Darwinism? It is Atheism. This does not mean, as before said, that Mr. Darwin himself and all who adopt his views are atheists; but it means that his theory is atheistic; that the exclusion of design from nature is, as Dr. [Asa] Gray says, tantamount to atheism. (Hodge 1874, 176–177)

It is a revealing statement. Plainly Hodge is not fond of "Darwinism," but his hostility is directed primarily at the exclusion of design from nature. Darwin's theory is consigned to the status of atheism—not by common descent per se, nor necessarily even by natural selection—but rather, as Hodge elsewhere explains at length, by Darwin's emphasis on natural selection as an undirected, nonteleological process. Darwin's account of evolution suggests that the appearance of humanity was the result of chance and was not foreordained. In this Hodge saw a contradiction with Scripture, which taught humanity's centrality to God's creative plan.

Today it is common to argue that evolution is simply God's way of creating. Hodge is contemptuous of that possibility:

> The second method of accounting for contrivances in nature admits that they were foreseen and purposed by God, and that He endowed matter with forces which He foresaw and intended should produce such results. But here his agency stops. He never interferes to guide the operation of physical causes. He does nothing to control the course of nature, or the events of history. On this theory it may be said, (1) That it is utterly inconsistent with the Scriptures. (2) It does not meet the religious and moral necessities of our nature. It renders prayer irrational and inoperative. It makes it vain for a man in any emergency to look to God for help. (3) It is inconsistent with obvious facts. We see around us innumerable evidences of the constant activity of mind. (ibid., 44–45)

Indeed, as adamant as he is that Darwinism is in conflict with Scripture, Hodge does not at all sound like a modern YEC. The first chapter of Genesis plays no role in his argument.

Hodge spoke for many. Some in the Protestant community were more willing than he to countenance Darwin's ideas, but with regard to undirected natural selection he carried the day. Historian Frederick Gregory summarizes the situation:

> Unquestionably, the attempt to reconcile evolution and Christianity depended on a rejection of natural selection as the mechanism of evolution. A few

> writers, for example Asa Gray and George Frederick Wright, claimed that natural selection was not incompatible with a divinely ordered creation, but after Hodge, theologians for the most part abandoned the attempt to reconcile natural selection and design.[6] (Gregory 1986, 383)

The Catholic Church responded along similar lines. Historians Mariano Artigas, Thomas Glick, and Rafael Martínez write:

> Evolutionism was viewed by many Catholic theologians as a materialist and agnostic ideology based on a scientific theory that had no serious foundation. This ideology seemed opposed to Christian doctrine on the Bible, on Creation and divine action in the world, and on human beings. There was a consensus among theologians about Catholic doctrine, including aspects that, without expressly being dogmas of faith, were held to be closely related to them, and evolution, as it was presented by its most ardent proponents, certainly clashed with dogmas of faith and with other positions generally held by theologians. (Artigas et al. 2006, 279–280)

This attitude prevailed also within the Congregation of the Index, charged with the responsibility of deciding if a theologically suspect book should be formally condemned and banned: "The archival documents make clear that, in the internal deliberations of the Congregation of the Index, there was an almost general consensus for rejecting evolutionism, above all with respect to the origin of the human body" (ibid., 274).

The Church hierarchy, still haunted by its poor treatment of Galileo, was reluctant to take a clear, public stand against Darwinism. The closest they came was in their decision to ban a book by Raffaello Caverni, a priest who defended a circumscribed version of evolution. It is interesting that Caverni explicitly excluded humans from the evolutionary process and rejected any nonteleological version of the theory (ibid., 35). This was still too much for the Congregation of the Index. Given the eerie consonance between their decision and issues in the modern debate, it is worth quoting at length:

> Until now the Holy See has rendered no decision on the system mentioned. Therefore, if Caverni's work is condemned, as it should be, Darwinism would be indirectly condemned. Surely there would be cries against this decision; the example of Galileo would be held up; it will be said that this Holy Congregation is not competent to emit judgments on physiological and ontological doctrines or theories of change. But we should not focus on this probable clamor. With his system, Darwin destroys the bases of revelation and openly teaches pantheism and an abject materialism. Thus, an indirect condemnation of Darwin is not only useful, but even necessary, together with that of Caverni, his defender and propagator among Italian youth.

> No less reproachable are Caverni's canons of scriptural exegesis . . . inasmuch as he seeks to limit the divine inspiration of the Holy Scripture to revealed dogmas and to morality, leaving unaffected by infallibility everything that sacred writers might teach with respect to the natural sciences. This exegetical system has recently been reproached by this Holy Congregation. . . . (ibid., 47)

We should note that, then as now, the Church authorities were generally more conservative and even fanatical than lay Catholics. We should not take their rather harsh reaction to Darwinism as necessarily typical of Catholics generally. That the Congregation of the Index had so much material to consider shows that at least some Catholics were more flexible than they. Still, the basic battle lines seem clear. Any notion that human beings are the result of a nontelelogical, fully natural process is contrary to Scripture and to Church teaching. Common descent among animals was less problematic, except to the extent that such thinking led naturally to an inclusion of humans in the process.

Mostly absent from the religious anti-evolutionism of this time was any concern for the age of the Earth or Noah's flood. When people spoke of conflicts between evolution and Scripture, they usually were not thinking of Genesis 1. Historian Ronald Numbers writes, "But in spite of the overriding biblical concerns of the leading clerical critics of evolution, virtually none of them insisted on compressing the history of life on earth into a mere six thousand years or invoked the Noachian deluge to explain the fossil record" (Numbers 2006, 26). The perceived conflicts largely revolved around the story of Adam and Eve, since the direct creation of the first couple by God seemed entirely at odds with notions of humanity's gradual appearance from eons of nonteleological natural selection. The centrality of the events in Eden to the Bible's broad story of sin and redemption, undermined by Darwin, was a further point of tension.

Evolution was hardly the only assault on the Bible's authority during this time. Starting in the late eighteenth century and continuing throughout the nineteenth, a group of predominantly German scholars had taken to studying the biblical texts from the standpoint of literary analysis. Among the major findings of this school of "higher criticism" was that the early books of the Bible were not authored by Moses, or by any single author for that matter. The integrity of other books was likewise called into question.

Seeking to push back against this challenge, a group of American Protestants published, between 1910 and 1914, a series of ninety essays known collectively as the "Fundamentals." The assembled authors were perturbed about many things, and the higher criticism loomed larger in their thoughts than did Darwinism. Lesser menaces like Mormonism and Roman Catholicism also came in for abuse.

The leitmotif of the entire series was aptly stated by Dyson Hague, a prominent evangelical cleric. He wrote:

> The present day liberal theology may be traced to two streams of influence: First, the influence of German rationalism, preeminently the Ritschlian theology, and the critical theories of Wellhausen, Kuenen and their school. Second, the widespread acceptance of the theory of evolution.
>
> To the first may be traced the free and easy way of the modernists of dealing with the Scriptures; and to the second, the revolutionized attitude of theologians with regard to sin, its source, its penalty, and its atonement. (Hague 2008, 91)

He later wrote, "The doctrine of evolution has washed out of the Bible the existence of such a man as Adam, and biology has taught that death is not due to sin" (ibid., 92). And later still:

> If man is God, there can be no guilt; and if there was no fall, and if it is the rise, not the fall of man with which the study of history makes us acquainted, there is, of course, no need for redemption; and if there is no need for redemption, there could, of course, be no ransom, or Redeemer, and an atonement is theologically and philosophically absurd. If there is no special creation, and man is a mere evolution from some frog or horse or anthropoid, why, of course, there can be no talk of atonement. (ibid., 93)

The Fundamentals contained three essays devoted specifically to the subject of evolution. Their titles leave little question regarding the viewpoints of their authors: "The Decadence of Darwinism," "The Passing of Evolution," and "Evolutionism in the Pulpit." This last essay, authored by an anonymous "Occupant of the Pew," contains passages such as this:

> Still more remarkable was the fact that so many theologians and Christian ministers adopted the new philosophy and were so ready to give up large portions of Holy Scripture because they could not be reconciled with it; inventing, as a salve to conscience, the doctrine that "the Bible was not intended to teach science," one of those half-truths that are more misleading than a down-right untruth.
>
> In this way the story of creation as given in Genesis was set aside, and the whole book discredited. As Christ could not by any logical possibility be made a product of evolution without an absolute denial of His supernatural birth and His Divine claims, and the new birth, or creation, for man in Him was open to the same objection, these truths were either obscured, minimized, or totally neglected and even denied. To such lengths were some of the sworn "defenders of the faith once delivered to the saints" ready to go in order to avoid being considered as hopelessly "un-scientific" and "behind the times in scholarship." (Anonymous 2008, 88–89)

Notice, though, that we still find in these essays no insistence that the Earth is young or that Noah's flood is relevant to geology. Most of the contributors were old-Earth creationists, meaning that they accepted the Earth's antiquity but rejected evolution.[7] In those essays that address evolution, the sticking point is primarily the lack of teleology in Darwin's theory. We even find statements such as this, from geologist George Frederick Wright: "Indeed, if it should be proved that species have developed from others of a lower order, as varieties are supposed to have done, it would strengthen rather than weaken the standard argument from design" (Wright 2008, 77).

Theologian James Orr likewise showed some flexibility. He reiterated the familiar objections to the nonteleological character of Darwin's theory and to the notion of man's gradual, as opposed to sudden, emergence. But he was willing to accept a degree of genetic relatedness among animals. With regard to a properly circumscribed and limited notion of evolution he wrote: "'Evolution,' in short, is coming to be recognized as but a new name for 'creation,' only that the creative power now works from *within*, instead of, as in the old conception, in an *external*, plastic fashion. It is, however, creation none the less." (Orr 2008, 346)"

There are three broad conclusions to be drawn from this regrettably brief history. First, the religious response to Darwin was not monolithic. Different aspects of evolution provoked different levels of religious opposition, with common descent, especially if it excluded humans, being easier to countenance than natural selection. Second, almost none of the opposition revolved around Genesis 1.

The third point, however, is that very few of the religious intellectuals of the time were prepared to accept evolution "full strength," as it were. Even those most willing to defend Darwin generally carved out an exception for humanity, or insisted that the variations on which selection operated were not random, or made other concessions toward keeping God directly involved.

This poses a problem, since the version of evolution these gentlemen feared is precisely the one that has triumphed today. No modern thinker could deny that humans are as much the result of evolution as any other species without being thought a creationist. Natural selection, in precisely the nonteleological form proffered by Darwin, holds pride of place among evolutionary mechanisms. It is logically possible that God is manipulating the genes in ways we cannot detect empirically, but the theory works quite well without such extravagances.

From Darwin right through to the present, substantial numbers of Christians have had serious reservations about evolution. It is not at all clear their concerns are unreasonable.

15

Intelligent Design vs. Young-Earth Creationism

Though I did not appreciate it at the time, my introduction to the difference between evolution and creation came when I was ten years old. I was driving with my father to his office, which involved traversing a small, narrow road flanked by farms. The narrow road eventually intersected a highway. We wanted to go straight through, but that was impossible because the continuation of our road was not directly across from us. Instead it was offset by a short distance. Starting from a full stop, we had to make a ninety-degree right turn onto the highway and then almost immediately make a ninety-degree left turn to continue on our way. Since there was always a lot of traffic, this was a dangerous and disconcerting maneuver.

I asked my father why the roads were designed in this convoluted way. He smiled and told me that these roads had not been planned at all. Actually, the highway was there first. Small towns gradually appeared on either side and each decided independently that it needed a connection. That is why the roads did not line up.[1]

This is an example of what paleontologist Stephen Jay Gould referred to as the "senseless signs of history." Viewed as the result of a city planner's design this intersection was absurd, but understood in the context of its history it made perfect sense. Gould applied this idea to evolutionary theory:

> Our textbooks like to illustrate evolution with examples of optimal design—nearly perfect mimicry of a dead leaf by a butterfly or of a poisonous species by a palatable relative. But ideal design is a lousy argument for evolution, for it mimics the postulated action of an omnipotent creator. Odd arrangements and funny solutions are the proof of evolution—paths that a sensible God would never tread but that a natural process, constrained by history, follows perforce.[2] (Gould 1980, 20–21)

These senseless signs of history are ubiquitous in nature. From a distance it is easy to marvel at the intricate design of the living world, but from up close things look considerably less marvelous. Everywhere we find desperate kludges and cobbled-together contraptions. Gould's observation cannot be dismissed as merely academic. We experience the cost of our evolutionary history in our weak lower backs, in endlessly rupturing appendixes, and in the wisdom teeth that often must be surgically removed. There are countless other examples in living organisms. None of this speaks well for any hypothetical intelligent designer.[3]

We have already noted that creationists carry with them a gallery of stock examples of biological complexity. They take such things to represent the defeat of evolutionary theory. The reality is precisely the reverse: when you really take the time to study biological complexity, it is evolution that comes to seem obvious. In complex system after complex system, biologists have found the senseless signs of history.[4]

I was mulling this over during the long drive from my home in Virginia to the Darwin vs. Design conference. This conference shall serve as an occasion to become familiar with intelligent design (ID). Unlike young-Earth creationists, the leading ID proponents seldom refer to Genesis, Noah's flood, or the age of the Earth. Most accept the Earth's great antiquity, and a few even allow room for the common descent of organisms. Officially the ID movement takes no stand on such questions.

Their biological fire is focused almost entirely on the explanatory sufficiency of natural selection. The same complex systems that seem to scientists to show clear evidence of evolutionary history are used by ID proponents as proof of design. No fully naturalistic process, they argue, can produce a functional machine as complex as the human eye or immune system. They assert that functional arrangements of parts are always evidence of intention and purpose.

We might suspect that the intentions and purposes manifested in nature are God's, but ID folks are adamant that this conclusion goes beyond their scientific theorizing. They treat "intelligence" as a particular sort of cause at work in the universe, one that can explain features beyond the reach of unaided nature. For example, erosion and weathering can account for the patterns of cracks and grooves on the side of a normal mountain, but they certainly cannot explain the faces on Mt. Rushmore. We must refer to the actions of intelligent agents for that and likewise, they argue, for certain aspects of the living world. If we are thereby forced to hypothesize the existence of powerful, nonhuman intelligences, then so be it.

The superficial sophistication of ID arguments, coupled with its lack of overt religiosity, has persuaded some that ID should not be considered creationism at all. Most scientists blush at the naivete of that view, urging that sometimes one must look beneath the surface. My own view is that the similarities between ID and creationism are far more significant than the differences, but let us take a closer look.

Several years ago I joined roughly thirty other people at an ID event in Washington D.C. After the formal presentation a young man approached me. We struck up a conversation, which was quickly joined by a young married couple. At some point I made a casual remark about creationism. To judge from their reactions, I had just committed a grave faux pas. They looked disgusted. My original conversation partner referred to creationism as "crap." The woman was even more blunt, describing creationists as "idiots" and "Bible-thumpers." It became clear that while all three were Christians, none had any respect for fundamentalism. This reaction has been common in my experience. Among ID supporters it is often believed that YECs, with their poor scientific arguments and extreme fundamentalism, have brought anti-evolutionism into unwarranted disrepute.[5]

The YECs are happy to reciprocate the ambivalence.

At the Creation Mega-Conference, I attended a talk entitled "The Intelligent Design Movement: How Intelligent Is It?" The speaker was Georgia Purdom, who holds a Ph.D. in molecular genetics from Ohio State University. Her verdict was mixed. She certainly agreed with the substance of the ID arguments and appreciated their success in gaining publicity for anti-evolutionary ideas. She fretted, though, that their unwillingness to identify the designer leads people only to a vague deism. "ID doesn't lead people to Christ," she protested. She had other objections as well, the most interesting of which was related to the problem of evil. Given the many examples of poor and malicious design in nature, the designer hypothesized by ID folks seems clumsy and unconcerned with the suffering of His creatures. If they refuse to incorporate the Bible into their science, then how are they to explain this?[6]

A more flamboyant version of the same criticism comes from Henry Morris (1999), one of the founders of modern young-Earth creationism. Referring to ID's arguments he writes, "On the other hand, a goodly number of atheists may convert to pantheism through such arguments. The various ethnic religions (Hinduism, etc.) all accommodate design, and so do the modern 'New Age' cults and movements."

Though there are certainly some people who maintain ties with both camps, the differences between ID and young-Earth creationism are real. At times the hostility of each toward the other rivals that of both toward evolutionists. There are genuine conflicts between the two camps, both in terms of general religious commitments and specific scientific opinions. That acknowledged, the proper analogy for their relationship is that of different dialects of the same language. This becomes clear from a consideration of the legal history of creationism, to which the remainder of this chapter is devoted.

It was bad enough when evolution was just an esoteric theory among scientists. By the dawn of the twentieth century, it had found its way into public school science curricula (Larson 1989 7–27). Legal attempts to rectify that situation soon followed.[7]

The first tactic was also the most direct: Laws were passed prohibiting the teaching of evolution. In Tennessee, in 1925, this took the form of the Butler Act, the relevant portion of which read:

> That it shall be unlawful for any teacher in any of the Universities, Normals and all the other public schools of the State which are supported in whole or in part by the public school funds of the State, to teach any theory that denies the Story of the Divine Creation of man as taught in the Bible, and to teach instead that man has descended from a lower order of animals. (Tenn. Code Ann. Title 49 [Education], 1922 [1925]

Later that year a substitute biology teacher named John Scopes agreed to be the defendant in a case testing the constitutionality of the law. The ensuing trial became a media sensation (de Camp 1968; Larson 1997). Scopes, inevitably, was convicted. He was fined one hundred dollars.

Two years later the case came before the Tennessee Supreme Court. The court's majority found, rather implausibly given the law's explicit reference to the Bible, that the Butler Act did not give preference to one religious group over another. The majority wrote:

> We are not able to see how the prohibition of teaching the theory that man has descended from a lower order of animals gives preference to any religious establishment or mode of worship. So far as we know, there is not a religious establishment or organized body that has in its creed or confession of faith any article denying or affirming such a theory . . . Since this cause has been pending in this court, we have been favored, in addition to briefs of counsel and various *amici curiae,* with a multitude of resolutions, addresses, and communications from scientific bodies, religious factions, and individuals giving us the benefit of their views upon the theory of evolution. Examination of these contributions indicates that Protestants, Catholics and Jews are divided among themselves in their beliefs, and that there is no unanimity among the members of any religious establishment as to this subject.[8] (*Scopes v. State of Tennessee*, 154 Tenn. 105 [1927])

The court also threw out Scopes's conviction on a procedural technicality (the fine should have been levied by the jury and not the judge). So ended the case.

Given the obvious parochiality of the court's decision—in which a law that makes the Bible the measure of educational legality was not found to give preference to one religious group over another, and in which the diversity of opinion among "Protestants, Catholics and Jews" was taken as representative of religious opinion generally—we should mention that it was only the religious liberty clauses in the Tennessee constitution that were under consideration. The "establishment clause" of the First Amendment to the United States Constitution, which says, "Congress shall make no law respecting an establishment of religion, or prohibiting the free exercise thereof," was at that time construed as restricting the federal government only. There were federal issues at stake in the Scopes trial, but these involved questions of due process and free speech.

Things changed with the 1947 decision by the United States Supreme Court in the case *Everson v. Board of Education*. The Court "incorporated" the establishment clause to state laws, meaning that the states had to abide by its restrictions no less than the federal government (Irons 2007). Aggrieved parties in these sorts of educational disputes could now seek relief in the federal courts.

In 1968, the United States Supreme Court, in the case *Epperson v. Arkansas*, struck down a 1928 law modeled after Tennessee's Butler Act. The decision was unanimous. The Court wrote:

> In the present case, there can be no doubt that Arkansas has sought to prevent its teachers from discussing the theory of evolution because it is contrary to the belief of some that the Book of Genesis must be the exclusive source of doctrine as to the origin of man. No suggestion has been made that Arkansas' law may be justified by considerations of state policy other than the religious views of some of its citizens. It is clear that fundamentalist sectarian conviction was and is the law's reason for existence. (*Epperson v. Arkansas*, 393 U.S. 97 [1968])

We should note that the Arkansas law in question did not mention the Bible specifically, but this was insufficient to render the law constitutional.

Since it was now impossible to outlaw evolution, creationists changed tactics. In 1961, Henry Morris and John Whitcomb Jr., respectively a hydraulic engineer and a theologian, published *The Genesis Flood: The Biblical Record and Its Scientific Implications*. They argued that geology had taken a tragic wrong turn when it discarded notions of a global deluge and a relatively young earth.

This inspired the invention of "creation science," which was put forth as a valid scientific alternative to evolution. Several states then passed laws mandating the teaching of creation science in any school where evolution was discussed. These were known as "equal time" laws. One such law was passed in Arkansas in 1981 and included this:

> Creation-science includes the scientific evidences and related inferences that indicate: (1) Sudden creation of the universe, energy and life from nothing; (2) The insufficiency of mutation and natural selection in bringing about development of all living kinds from a single organism; (3) Changes only within fixed limits of originally created kinds of plants and animals; (4) Separate ancestry for man and apes; (5) Explanation of the earth's geology by catastrophism, including the occurrence of a worldwide flood; and (6) A relatively recent inception of the earth and living kinds. (Ark. Stat. Ann. Title 80-1663, et. seq. [1981 Supp.])

The inevitable trial, *McLean v. Arkansas Board of Education,* took place later that year in a United States federal court, but the legal issues were now more complex. Creationists argued that theirs was an entirely scientific theory that happened to have religious implications. They claimed that the conclusions of creationism were those best justified by the evidence, their consonance with the Bible being merely coincidental. Since nothing in the Constitution barred the teaching of bad science, it was insufficient to argue that most scientists disagreed. Instead it had to be shown that creation science was not science at all. Calling it so was strictly a ploy for inserting a religious agenda into public schools.

Toward that end the plaintiffs presented scientific and philosophical testimony establishing that evolution was science and creationism was not. Persuaded by this testimony, Judge William Overton ruled that creationism's scientific pretensions were a sham and found the law to be in violation of the First and Fourteenth amendments to the Constitution. He wrote:

> The two model approach of the creationists is simply a contrived dualism which has no scientific factual basis or legitimate educational purpose . . . In addition to the fallacious pedagogy of the two model approach, Section 4(a) lacks legitimate educational value because "creation-science" as defined in that section simply is not science. (*McLean v. Arkansas*, 529 F. Supp. 1255, 1258 [E. D. Ark. 1982])

In the 1987 case *Edwards v. Aguillard*, the United States Supreme Court struck down a Louisiana law effectively identical to the one in Arkansas. The reasoning employed by the Court closely paralleled that of Judge Overton.[9]

This was the legal backdrop against which ID emerged. Intelligent-design proponents follow YECs in attempting a two-model approach, putting evolution on one side and their own views on the other. However, their creationist alternative no longer makes reference to anything recognizable from the book of Genesis. It claims instead that under certain circumstances the involvement of an intelligent agent in some event or physical structure can be proved rigorously. Among these structures are certain complex biological adaptations. As ID proponents see it, they have made significant conceptual advances since the natural theology of the nineteenth century. Whereas Paley argued mostly at an intuitive level, they say, they can bring mathematical and scientific rigor to these questions (Dembski 2004, 36–37).

Note that the "intelligent agent" they see lurking behind nature's contraptions is never explicitly identified as God—for all their theorizing can discern, we might be talking about awesomely powerful, but fully naturalistic aliens (Dembski 2004, 188) or even time-traveling cell biologists (Behe 1996, 249). That aside, ID proponents are equally adamant that their approach will liberate science from the materialistic doldrums in which it has foolishly mired itself.

Things looked promising for ID during the 1990s and early 2000s. A series of successful books laid out the ID case with a level of sophistication well beyond what the YECs could muster. Phillip Johnson, a law professor at the University of California at Berkeley, proved to be an effective strategist and organizer for the movement. His anti-evolution book *Darwin on Trial*, written as it was by a Harvard-educated lawyer from a university known for political liberalism, created something of a sensation. Subsequent works by Michael Behe and William Dembski, respectively a biochemist and a mathematician, brought novel arguments and intellectual heft. Major media outlets such as the *New York Times* and the *Washington Post* granted respectful coverage to the movement. It seemed as though creationism might finally have reached the mainstream.[10]

However, scientists and philosophers were distinctly unimpressed with ID's arguments. They found them inadequate on their merits and readily recognizable as variations on the standard creationist canon (Forrest and Gross 2004; Miller 1999, 2008; Pennock 1999, 2001; Perakh 2003; Petto and Godfrey 2007; Pigliucci 2002; Sarkar 2007, Shanks 2004; Young and Edis 2004). Still, as ID gradually pushed out traditional creationism and assumed an ever more prominent position in the public conversation, the question lingered: Was this a constitutionally acceptable version of creationism?

In November 2004, the Dover Area School Board in Pennsylvania voted to require its science teachers to read a statement to their classes. The statement emphasized that evolution was just a theory, presented ID as an alternative scientific theory, and referred students to a pro-ID textbook called *Of Pandas and People*. A group of parents and teachers in the affected area filed suit, and thus began the long-awaited legal test of ID.

The federal case, *Kitzmiller v. Dover*, played out over several weeks in late 2005. Judge John E. Jones III heard testimony both for and against the legitimacy of ID. Once again philosophers were called to demarcate science from nonscience, and they excluded ID from the area of legitimate science. Scientists testified to the bankruptcy of ID's scientific arguments. Members of the school board testified to the explicitly religious motivations of the policy, as expressed openly during board meetings.

Especially important was the testimony relating to the history of ID. It was not difficult for the plaintiffs to establish that ID was simply an outgrowth of creationism. Noteworthy in this regard was a discovery made by Barbara Forrest, a philosopher and expert witness for the plaintiffs. The book *Of Pandas and People*, mentioned in the Dover statement, had gone through multiple drafts. The earliest of these were written with creation science in mind and were meant to implement the "two-model" approach. After it became clear that creation science would not survive a legal challenge, the book was transformed into one promoting ID. In a comparison of the texts, Forrest found numerous passages that were word-for-word identical, but with words like "creationism" and "creator" replaced with "intelligent design" and "designer." Nicholas Matzke, a consultant for the plaintiffs' legal team, describes an especially damaging example:

> As if this evidence wasn't remarkable enough, while examining the drafts, Forrest discovered a peculiar sentence in the second 1987 draft of *Pandas*. "Evolutionists

> think the former is correct, cdesign proponentsists [*sic*] accept the latter view." Apparently, an editor was replacing the word "creationists" with the phrase "design proponents," but in the course of this tedious procedure the "c" and "ists" of "creationists" were accidentally left in the text. (Matzke 2009, 383)

This was one plank in a much larger case establishing that ID was just a legally savvy version of creationism. It is also a useful symbol of how things went for the ID side during the trial. No aspect of their case survived the skillful cross-examination of the plaintiffs' lawyers.[11] Judge Jones eventually issued a lengthy opinion demolishing every one of ID's substantive claims. Regarding the question of whether ID was science, he wrote:

> After a searching review of the record and applicable case law, we find that while ID arguments may be true, a proposition on which the Court takes no position, ID is not science. We find that ID fails on three different levels, any one of which is sufficient to preclude a determination that ID is science. They are: (1) ID violates the centuries-old ground rules of science by invoking and permitting supernatural causation; (2) the argument of irreducible complexity, central to ID, employs the same flawed and illogical contrived dualism that doomed creation science in the 1980's; and (3) ID's negative attacks on evolution have been refuted by the scientific community. As we will discuss in more detail below, it is additionally important to note that ID has failed to gain acceptance in the scientific community, it has not generated peer-reviewed publications, nor has it been the subject of testing and research. (*Kitzmiller v. Dover*, 400 F. Supp. 2d 707 [M. D. Pa. 2005])

Which brings us back to our original question: Is it fair to treat ID as just a form of creationism?

Your answer to that question will depend greatly on what you consider creationism to be. If you see it primarily as a specific set of beliefs regarding the age of the earth, the sudden creation of life, and the reality of Noah's flood, then ID is not creationism. Believing that the complexity of life suggests an intelligent designer does not compel you to accept that the designer is the Christian God. It certainly does not compel you to specific interpretations of Genesis. These differences have resonance in practical terms, as shown by the mutual suspicion between the two sides.

I would suggest, however, that this is not the most useful way of understanding creationism. At its core, creationism is a cultural and political rebellion against a scientific theory that is believed to menace religion and morality. Seen in this way, ID is absolutely a form of creationism, one that was born from the failures of previous strategies. The substance of ID arguments is only superficially different from traditional creationism, while its morally outraged rhetoric is identical to it.

Not to mention the most important similarity of all: Neither has produced a tenable, scientific argument against evolution.

16

Rhetorical Legerdemain

As a coda to our discussion in chapter 15, we should note a further, immensely frustrating, point of contact between young-Earth creationism and ID: The leading representatives of both have proven themselves willing to engage in highly unsavory rhetorical practices. Previous chapters have shown YECs at work—ID proponents are no less capable of habitually quoting scientists out of context and caricaturing their ideas.

A first example: In his book *Darwin's Black Box*, Lehigh University biochemist and ID proponent Michael Behe includes a section entitled, "The Natives Are Restless," which consists of a series of quotations from various scientists meant to show their great dissatisfaction with current evolutionary theory. For instance, Behe writes:

> A raft of evolutionary biologists examining whole organisms wonder just how Darwinism can account for their observations. . . . Jerry Coyne, of the Department of Ecology and Evolution at the University of Chicago, arrives at an unanticipated verdict:
>
> > We conclude—unexpectedly—that there is little evidence for the neo-Darwinian view: its theoretical foundations and the experimental evidence supporting it are weak. [Behe 1996, 26, 29]

Replying to this in an essay for the *Boston Review*, Jerry Coyne remarked:

> Apparently I am one of those faint-hearted biologists who see the errors of Darwinism but cannot admit it. This was news to me. I am surely numbered among the more orthodox evolutionists, and hardly see our field as fatally flawed. . . . Our paper cast no doubt whatever on the existence of evolution or the ability of natural selection to explain adaptations. (Coyne 1997)

The basis for Coyne's surprise becomes clear upon placing his statement in its proper context. It comes from a paper co-written with biologist H. Allen Orr:

> Although a few biologists have suggested an evolutionary role for mutations of large effect, the Neo-Darwinian view has largely triumphed, and the question

> is considered so dead that few may know the evidence responsible for its demise. Here we review this evidence. We conclude—unexpectedly—that there is little evidence for the Neo-Darwinian view: its theoretical foundations and the experimental evidence supporting it are weak, and there is no doubt that mutations of large effect are sometimes important in adaptation. (Orr and Coyne 1992)

Notice that Behe presented only part of the sentence he quoted. The omitted clause provides essential context. Coyne and Orr were suggesting that evolutionary biologists had unnecessarily limited their explanatory options on the narrow question of the genetics of adaptations. They certainly were not addressing the broader questions in which Behe was interested.

A second example comes from ID proponent William Dembski, in his book *No Free Lunch,* as he makes claims about the Cambrian explosion. The Cambrian era is the period of Earth's history roughly four hundred to six hundred million years ago. It is in rocks of this age that we find some of the oldest fossils on Earth. (*Cambria* is the Latin name for the region known today as Wales, where the first such rocks were found.) The "Cambrian explosion" is a period of roughly fifty million years during which the first representatives of most animal phyla make their appearance. In geological terms, the appearance of varied, complex forms over so small a duration can be considered sudden, and superficially this seems to conflict with Darwin's emphasis on the gradual development of life. Creationists of all stripes have been keen to emphasize that fact. Dembski writes, "At the same time, discontinuities in the fossil record (preeminently in the Cambrian explosion) are more difficult to square with common descent" (Dembski 2002, 315). Dembski then directs the reader to an endnote, which provides a quotation from paleontologist Peter Ward. According to Dembski:

> Ward writes, "The seemingly sudden appearance of skeletonized life has been one of the most perplexing puzzles of the fossil record. How is it that animals as complex as trilobites and brachiopods could spring forth so suddenly, completely formed, without a trace of their ancestors in the underlying strata? If ever there was evidence suggesting Divine Creation, surely the Precambrian and Cambrian transition, known from numerous localities across the face of the earth, is it." Note that Ward is a well-known expert on ammonite fossils and does not favor a creation-based view.[1] (ibid., 372)

Dembski is quoting from Ward's book *On Methusaleh's Trail: Living Fossils and the Great Extinctions* (1991).

Ward's words take on quite a different cast when considered in context. Following the portion quoted by Dembski, Ward describes Darwin's own concerns about this issue. He also discusses various explanations offered by Darwin's contemporaries, and shows how they fared in the light of subsequent developments. Then, after several pages, Ward comes to more modern views:

> Until almost 1950 the absence of metazoan fossils older than Cambrian age continued to puzzle evolutionists and earth historians alike. Other than the remains of single-celled creatures and the matlike stromatolites, it did indeed look as if larger creatures had arisen with a swiftness that made a mockery of Darwin's theory of evolution. This notion was finally put to rest, however, by the discovery of the Ediacaran and Vendian fossil faunas of latest Precambrian age.[2] (Ward 1991, 35)

On the next page Ward continues:

> Intensive searching of strata immediately underlying the well-known basal Cambrian deposits in the years between 1950 and 1980 showed that the larger skeletonized fossils (such as the trilobites and brachiopods) that supposedly appeared so suddenly were in fact preceded by skeletonized forms so small as to be easily overlooked by the pioneering geologists. (ibid., 36)

He then summarizes his conclusion:

> The long-accepted theory of the sudden appearance of skeletal metazoans at the base of the Cambrian was incorrect: the basal Cambrian boundary marked only the first appearance of relatively large skeleton-bearing forms, such as the brachiopods and trilobites, rather than the first appearance of skeletonized metazoans. Darwin would have been satisfied. The fossil record bore out his conviction that the trilobites and brachiopods appeared only after a long period of ancestral forms.[3] (ibid., 35–37)

That is all marvelously unambiguous and puts to rest any notion that Ward's statement can properly be used to support an anti-evolution argument based on the Cambrian explosion. In referring to divine creation, he was not expressing his own view, or even the view of any particular modern paleontologist. He was instead describing—with a bit of a rhetorical flourish—how things appeared prior to modern discoveries beginning in the 1950s.

Such examples of misquotation could be multiplied easily (Pieret 2006, Rosenhouse 2002b), but we should also mention a few other rhetorical tricks. For example, ID proponents are fond of making bold, confident presentations of arguments that are entirely incorrect. Here is Phillip Johnson, who was largely responsible for founding the ID movement in the early 1990s:

> [Biologist Douglas Futuyma writes] "The gradual transition from therapsid reptiles to mammals is so abundantly documented by scores of species in every stage of transition that it is impossible to tell which therapsid species were the actual ancestors of modern mammals." But large numbers of eligible candidates are a plus only to the extent that they can be placed in a single line of descent that could conceivably lead from a particular reptile species to a particular early mammal descendant. The presence of similarities in many different species that are outside of any possible ancestral line only draws attention

> to the fact that skeletal similarities do not necessarily imply ancestry. The notion that mammals-in-general evolved from reptiles-in-general through a broad clump of diverse therapsid lines is not Darwinism. Darwinian transformation requires a single line of ancestral descent. (Johnson 1991, 76)

A scientist reading that would find it hard to fathom what Johnson could possibly be talking about. The theory that modern mammals evolved from ancient reptiles implies that there used to exist creatures with anatomies intermediate between the two groups. Futuyma noted that the fossil record proves that such creatures actually existed, in such numbers and stages of transition that any line drawn between the former and the latter is hopelessly arbitrary. It is mysterious how this wealth of fossil evidence counts against evolution.[4]

Still another trick is the use of casual slanders against the integrity of scientists. Here is Johnson again:

> If the evolutionary scientists were better informed or more scientific in their thinking, they would be asking about the origin of information. The materialists know this at some level, but they suppress their knowledge to protect their assumptions. (Johnson 2000, 167)

This charge of rampant intellectual dishonesty, tossed off without a trace of supporting evidence, is standard in ID literature. Notice also how the "evolutionary scientists" of the first sentence became the "materialists" of the second. Conflating evolution with materialism is likewise commonplace, and could be listed as a separate rhetorical trick.

Nor should we overlook "the rant" as a rhetorical device. This is when the ID proponent drops all pretense and tells us what he really thinks. We shall close with a paradigmatic example, from Jonathan Wells:

> Dogmatic Darwinists begin by imposing a narrow interpretation on the evidence and declaring it to be the only way to do science. Critics are then labeled unscientific; their articles are rejected by mainstream journals, whose editorial boards are dominated by the dogmatists; the critics are denied funding by government agencies, who send grant proposals to the dogmatists for "peer review"; and eventually the critics are hounded out of the scientific community altogether.
>
> In the process, evidence against the Darwinian view simply disappears, like witnesses against the Mob. Or the evidence is buried in specialized publications, where only a dedicated researcher can find it. Once critics have been silenced and counter-evidence has been buried, the dogmatists announce that there is no scientific debate about their theory, and no evidence against it. Using such tactics, defenders of Darwinian orthodoxy have managed to establish a near monopoly over research grants, faculty appointments, and peer-reviewed journals in the United States. (Wells 2000, 235–236)

Given the ubiquity of this level of invective, is it really so hard to understand why scientists are hostile towards ID?

A scientist confronting ID sees first a collection of scientific and philosophical claims that are flatly erroneous in all of their major points. Upon investigating further he notes much that smacks of propaganda, but far less in the way of measured and reasoned discussion. If he is at all prominent in areas of interest to creationists, he can expect to find his words misquoted, his ideas distorted, his theories caricatured, and his integrity impugned. This, mind you, from people endlessly claiming God's blessing for their project. If the scientist protests, say by pointing out the thinness of ID's arguments, he gets accused of fomenting atheist propaganda and of existing in a state of massive cognitive dissonance. He is called a dogmatist. He and his colleagues are likened to the Mafia.

And when they are done hurling their invective, distortions, and misquotations, the ID folks then turn around and accuse scientists of being arrogant.

Where I come from we call that chutzpah.

17

Conversion Stories

Conversion stories feature prominently in creationist rhetoric. Just as evangelical Christians delight in telling you of the moment they came to a saving faith in Jesus Christ, so, too, are creationists eager to tell you when they became aware of the inadequacies of Darwinian evolution. Rare is the anti-evolutionist who was not initially in thrall to Darwin, prior to having his eyes opened by a wise and patient Christian teacher.

An interesting representative of both genres at once is the following account from Hank Hanegraaf, a popular Christian writer and talk-radio host. After describing his upbringing in a Christian home, he writes:

> But there was a catch. To experience true peace, I had to be willing to submit my life to Christ rather than travel the road of life by the dictates of my own will. And this I was not willing to do. I did not want to be deprived of all the pleasures the world had to offer, so I chose *rebellion* instead of *repentance*. It was not hard for me to justify my decision. At the same time that I was reading Scripture in church, I was reading science in class. And it was in science that I thought I had discovered an avenue of escape from accountability. (Hanegraaf 1998, 4)

He persisted in his state of happy rebellion until this:

> One wet and windy January evening in 1979, however, three Christians paid me a visit. They reminded me that God loved me and had a transcendent purpose for my life. They claimed Jesus Christ could replace my emptiness with peace and the Bible was a road map that would guide me through the maze of life. They were so kind and thoughtful that I felt a tinge of guilt for responding to their compassion with sarcasm and condescension.
>
> I could not resist the urge to point out that science had long ago demonstrated that God was little more than an illusion, that the Bible was a book for narrow-minded obscurantists, and that humans and their religious notions were merely functions of evolutionary processes.

> My guests listened politely until I paused to catch my breath. Then one of them asked if I was open-minded enough to objectively consider the evidence on both sides of the origins debate so I could make an *informed* decision. They told me that the following week their church would be hosting a workshop in which evolution would be examined in the light of the empirical laws of science. With that they left. (ibid., 5)

You can probably guess how the story ends. Hanegraaf went to the church, heard what he took to be an honest presentation of the evidence, and never looked back.

I have lost count of the number of times I have heard this story. I have heard it thundered from podiums and presented calmly in personal conversations. The major themes are always the same: Acceptance of evolution is equated with rebellion against God. Christian missionaries are models of wisdom and patience, while the evolutionists are smug and arrogant. An honest and open-minded consideration of the evidence, not religious concerns, leads to a rejection of evolution.

These stories never include the logical next step. I refer to the part where, after hearing the church's version of the evidence, the protagonist then seeks out a reply from someone knowledgeable about evolution. The possibility that it is the anti-evolutionists who are presenting a skewed and biased version of things never seems to occur to them.

The standard conversion story is somewhat revised in ID presentations. Keen to downplay the religious aspects of their thinking, they usually emphasize the moment when they were forced out of their mainstream-following complacency (as opposed to the moment when they gave their lives to Jesus Christ). A typical example comes from Jonathan Wells:

> During my years as a physical science undergraduate and biology graduate student at the University of California, Berkeley, I believed almost everything I read in my textbooks. I knew that the books contained a few misprints and minor factual errors, and I was skeptical of philosophical claims that went beyond the evidence, but I thought that most of what I was being taught was substantially true.
>
> As I was finishing my Ph.D. in cell and developmental biology, however, I noticed that all of my textbooks dealing with evolutionary biology contained a blatant misrepresentation: Drawings of vertebrate embryos showing similarities that were supposed to be evidence for descent from a common ancestor. But as an embryologist I knew the drawings were false. Not only did they distort the embryos they purported to show, but they also omitted earlier stages in which the embryos look very different from each other.[1] (Wells 2000, xi)

I will not think you overly cynical if you find these stories a bit too perfect. To me they seem like the sorts of tales that get a little better every time they are told. In

Wells's case we can ground our suspicions in an essay he wrote for Sun Myung Moon's Unification Church. The essay was entitled "Darwinism: Why I Went for a Second PhD."

> He [Reverend Moon] also spoke out against the evils in the world; among them, he frequently criticized Darwin's theory that living things originated without God's purposeful, creative activity. My studies included modern theologians who took Darwinism for granted and thus saw no room for God's involvement in nature or history; in the process, they reinterpreted the fall, the incarnation, and even God as products of human imagination.
>
> Father's words, my studies, and my prayers convinced me that I should devote my life to destroying Darwinism, just as many of my fellow Unificationists had already devoted their lives to destroying Marxism. When Father chose me (along with about a dozen other seminary graduates) to enter a Ph.D. program in 1978, I welcomed the opportunity to prepare myself for battle. . . .
>
> I eventually decided to join the fray by returning to graduate school in biology. I was convinced that embryology is the Achilles' heel of Darwinism; one cannot understand how organisms evolve unless one understands how they develop. In 1989, I entered a second Ph.D program, this time in biology, at the University of California at Berkeley. While there, I studied embryology and evolution. (Wells 1996)

These two accounts are plainly at odds. I will let the reader decide which he believes.

The "Darwin vs. Design" conference opened with two conversion stories. The first came from one of the conference organizers:

> I think probably everyone in this room has a different story that led them to be interested to come today . . . and I want to just tell you mine, briefly. I was, by training, a criminologist. I was a practicing investigator. . . . One day when I was on my way back to the station a radio call came out of a disturbance at an optometrist's office. . . . But I was close by and so I radioed in that I'd back up the uniformed officer that was responding. Well, as I walked into the door of that optometrist's office, I was shocked to see in front of me a fourteen-year-old boy bending over the body of the receptionist. He had a bloody knife in one hand and with the other he was rifling through her purse. . . . And on the way to the station I asked this young man why in the world would you do this? You got eight dollars out of her purse, and you killed this woman who had three children and a husband, and so you left the children without a mother and you've left this husband without a wife. . . . And he turned back to me and said, "I'm a juvenile, they're not going to do anything to me. And number two, it's survival of the fittest, man. It's survival of the fittest." And that started my own quest understanding what in the world is going on with our society.

The second came from Lee Strobel, who spoke next. He is a prominent Christian speaker and author, having been launched to fame with his 1998 book, *The Case for Christ*. This was followed by books such as *The Case for Faith*, *The Case for a Creator*, *The Case for Christmas*, *The Case for Easter*, and *The Case for the Real Jesus*, along with various student and children's editions of these same titles. He said:

> I want to tell you a bit about my story. It really goes back to the year 1970. At the time I was a high school freshman at Prospect High School in Mt. Prospect, Illinois, and I can take you back to the exact place that I was sitting when I lost all remnants of any belief in God that I had. It was in biology class, we were on the second floor, overlooking the asphalt parking lot, I was in the second row of seats from the window, third seat from the back. And why I remember this so distinctly is because the teacher got up that day and he described something that absolutely blew my mind!

The teacher told him about Stanley Miller's famous experiment, first described in 1953. Miller showed that many of the amino acids necessary for life could form spontaneously from simpler chemicals likely to have been present in the early atmosphere. It was the first major discovery in origins-of-life research. The importance of this experiment was not lost on young Strobel:

> A lightbulb went off in my head, because I realized instantly the implications of this. My conclusion was, "Well, God is out of a job." Because if you can demonstrate that the building blocks of life just occurred naturalistically as easy as that, then certainly there is no need for any outside influence or outside creator to be involved with the instigation of life in this world. God was out of a job.

Perhaps this seems implausibly melodramatic to you. Perhaps you are thinking that high school boys are rarely so affected by anything they hear in their classes, unless the teacher is very pretty. That is what I was thinking, at any rate, but I assure you the audience was riveted.

As you have no doubt anticipated, this was a set-up for Strobel's eventual emergence from darkness into light. His innate skepticism, he told us, led him to a career in journalism, and he eventually became the legal editor of the *Chicago Tribune*. Enter the wise and patient Christian teacher, this time played by his wife:

> I married a woman who was an agnostic, didn't really have an opinion one way or the other about the existence of God. I had a happy marriage until one day she came up to me and said, "You know, I've been checking things out for a while, and I've decided to become a follower of Jesus. I've decided to become a Christian." I thought, oh great, this is going to be the worst possible thing that could happen. But instead, through the succeeding months I began to see

> positive changes in her values and character, and so it intrigued me, and it encouraged me.

This, inevitably, was followed by the open-minded consideration of the available evidence:

> And I thought, I'm going to use my legal background, my journalism background, and I'm going to systematically investigate the evidence for Christianity and I'm going to start with the question, is there any evidence for the Creator? . . . And so I began to look at the scientific evidence, and see in which direction it pointed.

It was anticlimactic when he revealed the results of his sober and serious consideration of the evidence. "For me this was not faith versus science. This was science versus science. And I personally concluded that believing in Darwinism would require a leap of faith that I just was not willing to make."[2]

18

On Religious Experience

During a break in the conference, I started chatting with a fellow sitting near me. He told me he was an evangelical Christian, and I asked him how he had arrived at his beliefs. He replied that on several occasions he had so closely felt "the presence of God" (his phrase) that it became impossible for him to deny His existence.

In the years I have been doing this, I have had quite a few people tell me similar things. Many of my fellow skeptics and atheists are entirely dismissive of religious experiences, especially when they are used as evidence for God. For reasons I shall explain, I agree with them that such things have no apologetic value. That said, experiences with the power to so transform a person's life are not to be dismissed lightly.[1]

I have been a tournament chess player for over twenty years and have certainly had my share of interesting experiences. The position shown in figure 18.1 occurred a while back at the World Open, which any player could tell you is among the most important events on the American chess calendar. I was playing black.

A glance at the position reveals that I had sacked my queen for the inadequate compensation of a rook and knight, a defensive maneuver born of desperation and practical necessity. My opponent has just moved his queen from e8 to e4, placing her on an active central square and attacking my offside rook. I had been glumly pondering the position for several minutes, but no experienced player will be surprised that I mostly wasn't thinking about where to move my rook. Instead I was considering all the clever things I wished I had done thirty moves earlier.

Then I had the experience. I find it difficult to express in words, but it was as though I suddenly understood something that had formerly been opaque. People use expressions like "the clouds cleared" or "the light shone through," and I suppose they capture this experience as well as anything. There was no intermediate state where I sort of saw what was going on. I went, instantaneously, from despondency over defeat to certainty of victory.

I played 1. . . .g5+. White can escape the check with 2. Kh5, but this loses to 2. . . .Rxh3 mate. The only alternative is 2. fxg5, winning a second pawn. But now black has 2. . . .Rxh3+!, and suddenly my opponent realized he was about to lose

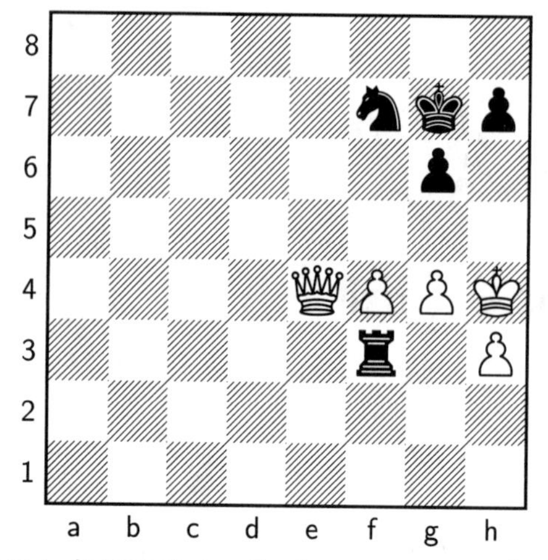

FIGURE 18.1 An interesting position from my career as a chessplayer.

the game. Play continued 3. Kxh3 Ng5+, forking the king and queen. From here the ending is trivial, and black will win with his one remaining pawn.[2]

I realize the non–chess players in the audience will not have understood that last paragraph, but any players reading this know *exactly* what I'm talking about. They are doubtless recalling their own, similar experiences. They know that the satisfaction of playing such moves transcends mere happiness. It is something to which you can always return. You can always recall the moment when you played a *really good move*. Perhaps you think this is exaggerated or overwrought, but that can only be because you have never had such an experience yourself. Within the community of chess players such things are understood and commonplace.

In his 1902 treatise *The Varieties of Religious Experience*, psychologist William James writes:

> One must have musical ears to know the value of a symphony; one must have been in love one's self to understand a lover's state of mind. Lacking the heart or ear, we cannot interpret the musician or the lover justly, and are even likely to consider him weak-minded or absurd. The mystic finds that most of us accord to his experiences an equally incompetent treatment. (James 1997, 300)

Having often been told, usually with condescension, that chess is just a game and therefore unworthy of such emotional commitment, I recognize much of what James describes.

My chess analogy can be pushed further. When I study the games of the great masters, I feel I am in the presence of abilities that far exceed my own. Folks like Bobby Fischer and Garry Kasparov seem to have a level of insight that would forever elude me, no matter how much time I put into study and practice. In this they can perhaps be likened to the great sages and prophets of the world's religions. They are figures possessed of singular abilities to which the rest of us can only aspire, whose examples provide inspiration for those not similarly gifted.

What I *can* do is engage in habits and disciplines that will increase my likelihood of having the requisite experiences. I can study the writings of the world champions. I can seek out a community of other chess players for support and guidance. I can immerse myself in a lifestyle devoted to understanding the game. I may never attain greatness, but I can certainly increase the richness of my own experiences at the board. The religious analogs are clear. Perhaps I will never have divine encounters to rival those of the great mystics, but through study and meditation I might enrich my own experiences of God. I can seek out a community of my coreligionists, people who understand my goals and can comfort me when I am frustrated.

Seen in this way, a sensitivity to religious realities can be viewed as a talent or ability found to varying degrees in different people. In his 1930 essay "The Argument from Religious Experience," philosopher C. D. Broad described it this way:

> Let us, then, compare tone-deaf persons to those who have no recognizable religious experiences at all; the ordinary followers of a religion to men who have some taste for music but can neither appreciate the more difficult kinds nor compose; highly religious men and saints to persons with an exceptionally fine ear for music who may yet be unable to compose it; and the founders of religions to great musical composers, such as Bach and Beethoven. (Broad 1998, 127)

Theologian Richard Swinburne expresses similar ideas, writing, "If some people do not have these experiences, that suggests that they are blind to religious realities—just as someone's inability to see colors does not show that the many of us who claim to see them are mistaken, only that he is color-blind" (Swinburne 1996, 115).

I have never had a religious experience. As I apparently lack the sensitivities described by Broad and Swinburne, I have made do instead with the many written accounts of such things, coupled with the scholarly literature devoted to their analysis. My chess analogy, hopefully not *too* strained, is my best attempt to liken anything I have found there to something in my own life.

It is often said that experience is the true wellspring of religion, with the churches and dogmas being secondary or even harmful. Writing in 1932, theologian Harry Emerson Fosdick gave eloquent voice to this idea:

> Let it stand as merely an indication of the major fact that multiplying numbers of people, when they think of religion, mean not a church, nor a system of theology, but a saving experience of inner spiritual devotion and daily spiritual power! . . .
>
> That, however, is the very nub of the theological and ecclesiastical problem as the school of thought from which this book comes apprehends it. The

> present churches and the present theologies have too little to do with this saving experience of genuine spiritual devotion and daily power. . . . Moreover, when the modern mind hears the creeds upon which many of the churches still insist, with all the corollaries brought out by controversy and urged as indispensable of religious truth—old cosmologies, doctrines of Biblical infallibility, miracles like virgin birth or physical resurrection—the reaction is not simply incredulity, although incredulity is undoubtedly emphatic—but wonder as to what such things have to do with religion. (Fosdick 1932, 21)

Fosdick, well known in his time for liberal theological views, surely went farther than most in downplaying the role of dogmas and creeds. His general attitude, however, has long been a centerpiece of religious thought. A more recent formulation comes from theologian John Haught:

> In our everyday existence the love of another person matters more to us than almost anything else, but gathering the "evidence" for that love requires a leap of trust on our part, a wager that renders us vulnerable to their special kind of presence. The other person's love, moreover, captures us in such a way that we cannot connect with it at all if we try to control it intellectually. Again, would it be otherwise in the case of any conceivable encounter of human persons with an infinite love? (Haught 2008, 46)

You can imagine my frustration at reading such things. Other aspects of religion I am permitted to treat intellectually and come to some reasoned conclusion. The traditional arguments for God have been clearly formulated by philosophers, making it straightforward for me to see their faults and inadequacies. The Bible can today be read by anyone willing to invest the time, and I can find in it not the slightest reason for suspecting it has a divine origin. A variety of academic disciplines shed light on the persistence and ubiquity of religion in defiance of the frequently incredible character of its doctrines and theologies.

Now I am told that all this time I have been missing the point. Apparently a true appreciation for religion requires possessing a brain attuned to such things. Lacking that, I must simply trust what other people tell me about it. This has the benefit of insulating religion from many rationalist assaults, as noted by philosopher J. L. Mackie:

> Since the early nineteenth century, and particularly through Kant's influence, the traditional proofs of theistic doctrines have been widely rejected or abandoned. . . . Also, we have seen how the problem of evil poses a very awkward question for anyone who wants to assert, literally, the full traditional set of theistic doctrines. A widespread response to these difficulties has been a shift of emphasis away from proofs and even from doctrines of a metaphysical sort, and a growing reliance instead upon religious experience. (Mackie 1982, 177)

Philosopher Eugene Thomas Long aptly summarizes the danger in this approach: "Experience of the divine is not recognized universally by all human beings and if one eschews understanding and argument in favor of religious experience, there is the risk that religion can be defended only by authoritative appeals to self-authenticating experience or revelation and faith" (Long 1991, 120).

That people have experiences that could fairly be described as religious is scarcely in doubt. I need not move beyond my own circle of friends and acquaintances to encounter some who have come to faith in moments of sudden conversion. As someone who rejects the notion of a supernatural realm, I obviously do not believe such people are receiving communications from God. Still, so long as the effect on their lives seems generally positive, I am not inclined to examine things too closely.

I would be happy to leave it there but for the prevalence of religious experience in the literature of Christian apologetics. The argument is made that the mystical experiences of other people comprises positive evidence of God's existence. Typically the argument is made by analogy to sense perceptions. Swinburne writes:

> Now it is evident that, rightly or wrongly, it has seemed . . . to millions and millions of humans that at any rate once or twice in their lives they have been aware of God and his guidance. . . . They may be mistaken, but that is the way it has seemed to them. Now it is a basic principle of rationality, which I call the principle of credulity, that we ought to believe that things are as they seem to be . . . unless and until we have evidence that we are mistaken. . . . If it seems to me that I am seeing a table or hearing my friend's voice, I ought to believe this until evidence appears that I have been deceived. . . . Just as you must trust your five ordinary senses, so it is equally rational to trust your religious sense. (Swinburne 1996, 131–132)

To go from, "You should trust your own religious experiences," to, "Other people should trust your religious experiences," Swinburne invokes a second principle:

> Someone who seems to have an experience of God should believe that he does, unless evidence can be produced that he is mistaken. And it is another basic principle of rationality, which I call the principle of testimony, that those who do not have an experience of a certain type ought to believe any others when they say that they do—again—in the absence of evidence of deceit or delusion. If we could not in general trust what other people say about their experiences without checking them out in some way, our knowledge of history or geography or science would be almost nonexistent. In virtue of the principle of testimony, there become available to those of us who do not ourselves have religious experiences the reports of others who do, and to which, therefore, we can apply the principle of credulity. (ibid., 116)

And his conclusion?

> So in summary in the case of religious experiences, as in the case of all other experiences, the onus is on the sceptic to give reason for not believing what seems to be the case. The only way to defeat the claims of religious experience will be to show that the *strong* balance of evidence is that there is no God. In the absence of that strong balance, religious experience provides significant further evidence that there is a God. (ibid., 118)

As it happens, I *do* think the balance of evidence is strongly against the existence of God, so long as we are talking about an omnipotent God of infinite love who seeks communion with His creatures. That evidence comes in the form of massive suffering and evil, coupled with the inability of so many honest seekers to find any trace of Him. These are known, respectively, as the argument from evil and the argument from divine hiddenness, and I have found nothing in the theological literature to mitigate their force.[3]

That said, there is no need to base our skepticism on anything quite so contentious as *that*. Reading Swinburne's breezy rhetoric about principles of rationality and the burden of proof being upon skeptics makes one suspect he is making life entirely too easy for himself.[4]

The notion of a parallel between mystical perception and sense perception has a long history in Christian thought. Consider the views of John Wesley in the eighteenth century, here summarized by theologian Theodore Runyon:

> Spiritual sensation operates in strict analogy to physical sensation. Just as the physical senses do not provide knowledge of the empirical world except as they are acted upon and stimulated by sense data, the spiritual senses have no knowledge or acquaintance with God and spiritual reality except as the Divine Spirit acts upon them. Although given with our creation, the spiritual senses (because of the fall and human sin) remain obscured, unused, atrophied. It is as if they did not exist—until such a time as they are "quickened," activated by the life-giving Spirit, and the image of God is renewed in us. (Runyon 1991, 189)

Surely, though, there is an obvious point of disanalogy between the two. Our physical perceptions are subject to numerous checks and verifications, which seems not to be the case with mystical perceptions. Contrary to Swinburne, we do not simply trust our sense perceptions in the absence of corroborating evidence. We do so both because of the internal consistency of what our senses reveal, and because of the supporting testimony of other people. Bertrand Russell, in a 1948 radio debate with the Jesuit priest Frederick Copleston, expressed things well:

> I should reply to that line of argument that the whole argument from our own mental states to something outside us, is a very tricky affair. Even where we all admit its validity, we only feel justified in doing so, I think, because of the consensus of mankind. If there's a crowd in a room and there's a clock in a room,

> they can all see the clock. The fact that they can all see it tends to make them think that it's not an hallucination: whereas these religious experiences do tend to be very private. (Russell and Copleston 1986, 134; see also Martin 1952, 504)

The need for such verifications is made manifest by the various physical maladies that can cause us to misperceive our surroundings. If I believe I am seeing something that other competent observers fail to see, then serious doubt is cast on the correctness of my belief. Likewise, we would all acknowledge that people can be mistaken regarding their mystical perceptions. Those who commit grave crimes in the sincere belief that God has commanded them to do so are typically described, without hesitation, as mentally ill. Moreover, people often have conflicting religious experiences. It follows that at least some of them are misinterpreting things.

Given this reality, some mechanism for distinguishing the true mystical perceptions from the false ones is crucial if we are to admit religious experience as evidence for God. With sense perceptions, we have both the internal consistency of our own experiences (we hear the glass break at the same instant we see it smash into the ground) and the corroborating testimony of other observers (they claim to have seen the glass break at the same instant we did). The reality of frequently conflicting religious experiences shows there is no such consistency in this case. It is hard to imagine what verification mechanism is available.

Philosopher William Alston, a defender of religious experience, has some suggestions in this regard.

> The first thing to be said in reply is that there *are* tests for the accuracy of particular reports of mystical perception. Contemplative religious communities have compiled systematic manuals of such tests; and many of them are used informally by the laity. These include such things as (1) conformity with what would be expected on the basis of doctrines concerning the nature of God, (2) "fruits" of the experience as a stable inner peace and growth in spirituality, (3) a content of the experience that the person would not have developed on their own. The satisfaction of such conditions counts in favor of the veridicality of the experience, and their absence counts against it. (Alston 2004, 143)

The first two items are, presumably, an appeal to the sort of internal consistency I discussed previously. A religion might teach that ultimate reality is friendly and welcoming, and then people have experiences that are consistent with that premise. Doctrine and experience are thus mutually reinforcing, in perfect analogy with the consistent testimony of all our various senses.

This, however, fails to do justice either to the variety of religious doctrines or to the varieties of religious experience. Could not a Satanist argue that his religion teaches the hostility of ultimate reality, and that this teaching is confirmed by the many people driven to poor behavior by their mystical encounters? What reason can be given for thinking that a stable inner peace is the proper end result of a veridical religious experience? Alston's first two criteria are doubly arbitrary. They involve choosing first the subset of religious doctrines we find congenial, and then discarding the experiences that are contrary to them.[5]

There is a further problem. A person tends to interpret his experiences through the prism of his culture. Philosopher John Morrison Moore, writing in 1938, expressed the point well:

> Within any given culture people learn to find particular religious meanings in certain types of experience, while in another culture these experiences might not be regarded as religious at all or might have a very different religious significance.
>
> The pluralistic nature of religious experience implies also that no general answer can be given to the problem of the cognitive value or authority of religious experience. . . . It must be remembered, furthermore, that interpretations which seem to come from the experience itself may in reality have been furnished as a framework of the experience from outside it. It is misleading to think of an interpretation simply as "growing out of " an experience, for our categories and established modes of reaction are present before any particular experience, and condition the form which the experience itself takes. (Moore 1938, 186–187)

In a more recent discussion, philosopher Philip Kitcher argued along similar lines:

> A sense of the presence of a deity (or to other sorts of beings) is very common across the spectrum of the world's religions. Too common, in fact, to play any serious evidential role. Once you appreciate the widespread tendency of people to arrive at very different claims on the basis of experiences that seem to them both intense and mysterious, and see that their experiences are categorized in terms that derive from the religions with which the subjects are familiar—and to which they often subscribe—it is clear what is occurring. (Kitcher 2011, 7)

This seems entirely correct to me, and it seriously undermines the claim that doctrine and experience are mutually reinforcing. What we have is not a consilience of independent lines of evidence, but a situation in which prior immersion in a given culture influences the manner in which a mysterious experience is interpreted. This relates as well to Alston's third condition, for how do we know what a person would or would not devise on his own? If a person with no inkling of Christianity believed he had an experience of someone named Jesus and was subsequently able to speak intelligently about subtleties of the faith, then we might have strong evidence for communication with a divine reality. In practice, however, people seem to have experiences of entities familiar to them from the dominant religion in their culture (Antes 2002).[6]

We have not yet considered the most obvious disanalogy between religious experience and sense perception. I know that my eyes are for seeing and my ears are for hearing. But when we speak of religious perception we might wonder what it is, exactly, that is doing the perceiving. John Morrison Moore addresses this point with a bit of impatience:

> It is sometimes held that religious experience depends upon a special sense by which the spiritual is directly apprehended, but it is hardly possible to maintain this view with any precise meaning or scientific justification. There is a flavor of the occult about such a conception which does not comport with its use in serious analysis. The notion that spiritual or religious entities are apprehended by means of a sixth sense, or faculty of metempirical perception, is a relic of a debased conception of the spiritual as composed of some form of rarefied or ghostly extended substance. (Moore 1938, 169)

Modern neuroscience has made progress in elucidating the physical basis for religious experience. The literature on "neurotheology," in which one attempts to relate spiritual perceptions and religious experiences to brain states, is quite large and growing rapidly (McNamara 2006). I fear it would strain both the reader's patience and the author's competence to survey this literature, but at least one conclusion does seem justified. That religious experiences can be triggered by entirely physical causes, such as microseizures in the temporal lobes or the ingestion of certain drugs, is undeniable.

This does not prove that all religious experiences have such mundane causes, but it does suggest that at least some of them do. This reinforces the necessity for a means of distinguishing true religious experiences from false ones.

But philosopher Eric Reitan scoffs at the significance of such things:

> While some researchers claim to have found a region in the brain that, when stimulated in the right way, produces religious experiences, what does that show? If my visual cortex is stimulated in the right way, I'll experience bright light. Does that mean the experience of light is just a by-product of neural misfiring and there really is no sun? Of course not. (Reitan 2009, 154)

Of course, in keeping with my previous argument, I would note that our evidence that the Sun is real is based on the massive internal consistency between the reports of our own senses and those of others, and not on an isolated visual perception. More to the point, however, a naturalistic account of some phenomenon never completely rules out supernatural involvement. It simply makes the supernatural seem superfluous, and that is the case here. A complete understanding of Newtonian mechanics leaves open the possibility of gravity gremlins at the heart of it all. But the gremlin defender can certainly be accused of favoring an elaborate explanation when a simple one suffices. If religious experiences can be explained by simple, physical causes, then why should we go looking for complex, supernatural theories?

* * *

I spent the summer of 1994 engaged in mathematical research at Oklahoma State University. It was part of a program intended for undergraduate students interested in pursuing graduate studies. Six students and three professors participated, and one night we all had dinner at our supervisor's house.

His wife had made, among other things, a pot of chili. It was quite delicious, especially after weeks of cafeteria food. As we left the house later that night I told her, somewhat tactlessly, considering that our hosts were quite conventionally religious, that her chili was so good it was like a religious experience.

She smiled and said, "Well, I hope you have a better experience than *that* someday."

Perhaps someday I will. For now, though, that chili remains well in the lead.

19

Creation as Fishtank

The morning sessions at the Knoxville conference provided a survey of the proposed evidence for design.

In Lee Strobel's presentation, discussed briefly in chapter 17, two points were emphasized. The first was the classical "first cause" or "cosmological" argument, known to philosophers for centuries. In Strobel's version (based on the writings of philosopher William Lane Craig) we start from the premise that anything that began to exist must have had a cause. Since modern cosmology tells us that our universe came into existence with the Big Bang, we conclude that the universe has a cause for its existence. And since naturalistic causes are insufficient, it is claimed, we must turn instead to personal explanations. That is, to God. For His part, God did not begin to exist. He is eternal. That is why we are not compelled to find a cause for God's existence. So goes the argument.

Strobel's second claim involved the "fine-tuning" of the fundamental constants of the universe. An example is the ratio of the mass of a proton to that of an electron. If that ratio were even slightly different from what is observed then atoms could not form and life would be impossible. Such fine-tuning is a common phenomenon. There are various constants which, if altered slightly in isolation from all other such constants, would render the universe inhospitable to life. Moreover, there is nothing in current theorizing to explain why the constants have just the values they do.

Since our universe plainly contains life, the impression is given that we have somehow won a great lottery by getting just the right fundamental values to make life possible. It strains credulity to attribute this to luck, so we must invoke the wisdom of an intelligent designer. Again, so goes the argument.

Strobel is a talented propagandist, and his enthusiastic presentation suggested that only a very foolish person could reject such evidence. Few philosophers or scientists agree, seeing in both arguments just a mess of unsupported claims and vague assertions. The correctness of his first gambit, known in the trade as "the Kalām cosmological argument," hinges on difficult questions about causality, infinity and modern physics. When presented with proper philosophical rigor it is clear the argument depends on too many unsupportable assumptions to be convincing.[1]

Likewise for the second argument, which suffers from a variety of flaws. For one example, let us reconsider the lottery analogy. Winning a lottery is impressive when you are only permitted one ticket, but is less so when you get to play millions of times. The cosmic equivalent is the possibility that our universe is just one of an essentially infinite number of others. If we exist in such a "multiverse," then finding one with just the right constants for life is no more remarkable than winning a lottery once after playing millions of times. As it happens, the reality of multiple universes is strongly suggested by current theorizing in cosmology, to the point where it has been an entirely mainstream idea among physicists for decades. Certainly the existence of multiple universes is speculative, and we do not have direct evidence for their existence. But since the God hypothesis suffers from the same defects that hardly seems like much of a criticism.[2]

The professional literature discussing Strobel's arguments is vast and difficult, a fact that plays into something we discussed earlier. So many anti-evolution arguments resonate with people because they are based on intuitive, commonsensical notions: *Complex machines require designers. Natural forces tend to break things down. Anything that begins to exist has a cause for its existence. Cosmic fine-tuning is like winning a lottery.* All these are notions that are easy to explain and seem justified by our daily experiences.

But science is good at exposing the inadequacies of common sense, while philosophy is equally adept at showing that seemingly simple notions often become very complex when examined closely. It takes training and education to appreciate this; where training and education are lacking, it is easy for vague and superficial arguments to take hold. This is a large part of the reason creationists often do very well in debates. Strobel's arguments were easily presented in a few minutes and resonated with the audience. By contrast, a scientist trying to refute him properly would quickly be forced into dense and technical material.

Another striking aspect of Strobel's presentation was the extent to which he relied on arguments from ignorance. He was simply inserting God into those places where we lack the data to draw proper conclusions. Physicists have been ingenious at unraveling the history of the universe, right back to the first moments after the Big Bang—but then our data runs out, and we are left with fundamental mysteries that might never be resolved. Many of the usual concepts with which physicists deal, like causality, time, and matter—ideas which seem so simple and straightforward in our daily lives—become very confusing when pushed to the frontiers of existence.

It is precisely there, at our moment of maximum ignorance, that ID proponents insert their most audacious conclusion. Just when the only dignified response is a shrug of the shoulders and a mumbled "Who knows?" they speak with complete confidence. They have everything backward. When biologists summon forth the massive evidence for evolution, drawn from every relevant branch of the physical and life sciences, the ID folks fight them tooth and nail and accuse them of arrogance and dogmatism. When the evidence runs out and we really should be humble and circumspect in our conclusions, they claim instead to have established the most dramatic and momentous facts regarding the nature of existence.

There were two further presentations that morning, emphasizing arguments drawn from astronomy and biology. Then we broke for lunch, leading to the events at the Subway sub shop, recounted in chapter 2.

Especially interesting about the morning's presentations was the near-complete absence of anything a professional scientist would regard as a scientific argument. The talks were informed by science, certainly, and there was no shortage of facts (not all of them accurate) to consider. The problem was the means by which they proceeded from their facts to their conclusions.

Let me illustrate. If you ask a scientist why he accepts evolutionary theory, he will probably answer along these lines: "Evolution makes specific predictions that we can test against the natural world. It has implications for the fossil record, genetics, biochemistry, and numerous other disciplines. In each case we find an impressive consonance between expectation and reality. This mass of converging evidence persuades us that the theory is substantially correct."

This style of thinking embodies the central concerns of working scientists. In pondering nature they seek control and predictive accuracy far more than absolute truth. The hallmark of a scientific theory is that it has testable consequences. We must be able to distinguish data that supports the theory from data that does not, and both varieties must be such that they could, at least in principle, be observed in nature.

The morning's speakers did not proceed in that fashion. They did not put forth a clear theory of design, deduce its consequences, and then compare those consequences with the actual data. Instead they began with certain scientific facts and simply asserted, with very little supporting argument, that design was the best explanation. In his professional role, a scientist would simply shrug and say, "Draw whatever metaphysical conclusions you want, but there is nothing here remotely helpful to my research."

In describing their arguments as unscientific I am not necessarily saying they are unreasonable.

For an analogy, imagine that you set up a large aquarium in your living room. To liven up the environment for the fish, you place a small castle on the left side of the tank. The following day you return and discover the castle is now on the right side of the tank. How would you explain this? You would reason, surely, that you know enough about the tank's workings to conclude that nothing from inside can explain the castle's journey. You could devise very fanciful explanations, like the possibility that one of the fish is genetically engineered for super strength and was able to push the castle from one side to the other, but barring extraordinary evidence you would not long entertain such suggestions. Instead you would conclude simply that someone reached into the tank from outside and moved the castle.

Could the fish be justified in drawing the same conclusion? We might imagine that the more scientifically inclined fish have produced a decent catalog of things that happen within the tank, coupled, perhaps, with theories that account for the various regularities that they observe. But certain mysteries remain. From time to time flakes of food appear on the surface, but nothing in the catalog explains how that is possible. Sometimes the physical environment changes dramatically, as when the castle moved from one side of the tank to the other. More religious fish might conclude, on the basis of their persistent failure to explain these phenomena solely from within the tank, that there are entities able to reach into the tank from outside.

Scientist fish might object that their research is hardly complete. Perhaps some unknown chemical process causes food to coalesce from the water itself. The gravel at the base of the aquarium, he might argue, is still largely mysterious. He could point to all the fish who looked foolish for suggesting, before the mechanics of the filter had been worked out, that only intervention from beyond could explain the disappearance of waste products from the water.[3] At any rate, he could argue that invoking mysterious external entities is not helpful, so he and his fellow scientists should be allowed to continue their work in peace.

Religious fish, I would think, could readily assent to all of this. He could acknowledge the need for proper scientific humility and express willingness to hold his views contingent on the appearance of new evidence. For all of that, however, the desire to take a stand cannot forever be denied. Religious fish will note that nothing the scientists have discovered is even remotely able to explain the appearance of food or the movement of castles. Therefore, he will tentatively conclude the existence of agents from outside the tank. He will then go on his way, allowing scientist fish to continue investigating by his undeniably useful methods.

Everyone is happy! Scientist fish continues his work *over here*, while religious fish draws his conclusions *over there*. Good fences make good neighbors.

Of course, in my story it is religious fish who is right and scientist fish who looks foolish. From my perspective that is a serious point of disanalogy between my story and real life, but I do find the example illuminating. In principle I see nothing wrong in saying, with regard to some observed event: "We know enough about the relevant natural forces to conclude they are inadequate for explaining this phenomenon. Design by a powerful intelligence is the most reasonable conclusion at this time."[4] If one of the Great Pyramids of Egypt were suddenly to rise straight up from the ground, invert itself, and come to rest, perfectly balanced, on its apex, no one would hesitate to conclude that intelligent agents were at work. A scientist who argued that the inverted pyramid is actually evidence of a previously unsuspected natural mechanism would just look silly. The mechanics of large objects are quite well understood, thank you, and a natural process capable of such dexterous pyramid manipulation would surely have been discovered by now.

Which raises an important question. Why are the ID folks so insistent that their ideas be considered a part of science?

In modern science we find people of very different religious opinions coming together in a common enterprise. Some, myself included, believe that atheism is the view that fits most comfortably with science. Others prefer a vague spirituality or a theologically liberal version of a well-established religion. Still others are entirely traditional in their outlook. Different people draw different metaphysical conclusions, but those are conversations to have at night over drinks, not during the day in the lab.

Why, then, do the ID folks find it so hard to respect a division between science and religion?

There are two obvious answers, both of which capture something important. The first relates to the status granted to science in modern discourse. For many people, describing something as scientific is effectively equivalent to saying it is true. By contrast, outside of a specific religious community, no one can plausibly cite theological teachings as legitimate justification for a factual claim. The desire to describe ID as "scientific" is thus seen as an attempt to capitalize on the prestige that this term conveys.[5] The second answer relates to education. Anti-evolutionists of all stripes are entirely sincere in their belief that evolution has profoundly negative consequences for the spiritual lives of students. They want to mitigate this harm by teaching their contrary ideas. It is unconstitutional to teach religion, but science that happens to have religious implications is allowable. Describing ID as good science is thus a political strategy for inserting it into science curricula, thus making use of public resources to promote their particular religious ideas.

There is also a third explanation, less obvious and insufficiently appreciated in much anti-evolution writing. It is this explanation that we will examine in the next chapter.

20

Methodological Naturalism

Among ID supporters "science" is commonly understood to refer to the totality of all that is true about nature. That God created the world is certainly part of that truth. When critics say things like "ID simply is not science," the typical ID proponent is aghast. He replies: "But isn't the goal of science to learn the truth about the world? God is part of that truth! So how can God not be part of science?"

Philosophers of science have long discussed the "demarcation problem," by which they refer to the attempt to draw a clear line between science and other sorts of inquiry. In this regard it has been suggested that science must adhere to a principle of "methodological naturalism" (MN). It is claimed that though the line between science and nonscience is sometimes blurry, an invocation of supernatural entities nonetheless places you automatically on the wrong side of it. The term "methodological" indicates that naturalistic presumptions are a matter of scientific practice only. Methodological naturalism is distinguished from "philosophical naturalism," which is the metaphysical position that supernatural entities do not exist. Science so restricts itself, the argument continues, because of its emphasis on predictability and control. Supernatural entities do not lend themselves to empirical tests, you see.

What seems to scientists like a commonsense principle dictated by the nature of their enterprise looks to ID proponents like an attempt to rule out ID by definitional fiat. Philosopher Robert Pennock defends the principle against this charge:

> It is misleading for creationists to characterize science in general and to define evolution in particular as "godless." Evolutionary biology is godless in the same way that plumbing is godless. Evolutionary biology is no more or less based on a "dogmatic philosophy" of naturalism than are medicine and farming. Why should [ID proponent Phillip] Johnson and his allies find methodological naturalism so pernicious and threatening in the one context and not the others. . . . Surely it is unreasonable to complain of a "priesthood" of plumbers who only consider naturalistic explanations of stopped drains and do not consider the "alternative hypothesis" that the origin of the backed-up toilet was the design of an intervening malicious spirit. Would it not be bizarre to reintroduce theistic explanations in the agricultural sciences and

> have agronomists tell farmers that their crop failure is simply part of God's curse upon the land because of Adam's disobedience, or suggest that they consider the possibility that the Lord is punishing them for some moral offense and that it might not be fertilizer they need but contrition and repentance? (Pennock 1999, 282)

It is difficult to express precisely how obtuse this seems to the average ID proponent. Naturalism in plumbing is acceptable because no one has ever expected to find supernatural causes underlying a clogged drain. In this it stands in stark contrast to the origins of humanity, which has traditionally been assumed to have a supernatural component. No one's beliefs about their place in the cosmos are riding on the work that plumbers do. Plumbing does not impinge on fundamental questions about meaning, value, purpose, or morality. That is why naturalism in evolution is pernicious and threatening, while naturalism in plumbing is not.

Pennock was surely amusing himself with his little examples about agronomy and demon clogs, but the ID proponent will only see an insult to her intelligence. She will wonder why, after concluding that certain aspects of living organisms are best explained by invoking design, she is suddenly obliged to consider supernatural explanations in every facet of daily life. Her willingness to consider the supernatural in biology is born neither from a desire for simple explanations nor from a belief that supernatural influences are omnipresent. A conclusion of design must itself be based on strong evidence and argumentation, she would argue, and should not be invoked when adequate natural mechanisms can be found. The point, from her perspective, is simply that we should not forever rule the supernatural out of bounds, especially when naturalistic explanations have routinely proven themselves impotent.

Here is a further line of argument used in defense of methodological naturalism, again from Robert Pennock:

> If one were to find some phenomenon that appeared inexplicable according to some current theory one might be tempted to attribute it to the direct intervention of God, but a methodological principle that rules out appeal to supernatural powers prods one to look further for a natural explanation. And it is not merely because such persistence has proved successful in the past that science should want to consider this attitude.
>
> The scientists' appeal to supernatural agency in the face of a recalcitrant research problem would be as profoundly unsatisfying as the ancient Greek playwright's reliance upon the *deus ex machina* to extract his hero from a difficult predicament. (ibid., 293)

This, too, will look very odd to an ID proponent. She will reject the implication that her method is, "If the phenomenon looks inexplicable, then I will casually invoke supernatural design." Instead, she will argue that her method is, "If the phenomenon looks inexplicable in a specific, quantifiable way that is strongly suggestive of design, and if the phenomenon has consistently resisted naturalistic explanations despite extensive research, then supernatural design is likely to be the correct answer, and prodding oneself to look further for naturalistic explanations is not a good idea."

Her greatest astonishment would be reserved for Pennock's assertion that an explanation based on supernatural design would be "profoundly unsatisfying." Perhaps so, she might reply, if the only thing that satisfies you is a detailed naturalistic account of every phenomenon. But if the ID theorists are right, then they have made a truly momentous discovery about nature. It is hard to imagine anything *more* satisfying.

Moreover, it is unclear if considerations of what is and is not satisfying have any proper role in this discussion. Rather a lot of people find evolution to be entirely disquieting, but that is not generally considered an argument against it. Philosopher Alvin Plantinga expresses the point well:

> The claim that God has directly created life, for example, may be a science stopper; it does not follow that God *did not* directly create life. Obviously we have no guarantee that God has done everything by way of employing secondary causes, or in such a way as to encourage further scientific inquiry, or for our convenience as scientists, or for the benefit of the National Science Foundation. Clearly we cannot sensibly insist in advance that whatever we are confronted with is to be explained by something *else* God did; he must have done *some* things directly. It would be worth knowing, if possible, which things he *did* do directly; to know this would be an important part of a serious and profound knowledge of the universe. (Plantinga 2001b, 357)

This sort of definitional rhetoric plays right into the hands of ID propagandists. It provides them with an explanation for why, if evolution is the utter nonsense ID says it is, the lion's share of the scientific community persists in embracing it as a strong and vibrant theory. Young-Earth creationists argue that our sinful natures drive us to anti-God ideas, but that is too overtly religious for ID purposes. Instead ID proponents argue that scientists are required by the arbitrary rules of their enterprise to dismiss out of hand an explanation that invokes design. Phillip Johnson writes:

> Victory in the creation-evolution dispute therefore belongs to the party with the cultural authority to establish the ground rules that govern the discourse. If creation is admitted as a serious possibility, Darwinism cannot win, and if it is excluded a priori Darwinism cannot lose. (Johnson 1990, 18)

It is impossible to overstate the centrality of this claim to ID argumentation.

At a 2003 ID conference held near Kansas City, I was in the audience for a one-act play, "The Rule." The plot was recognizable as a standard conversion story. In the role of the put-upon-but-endlessly-patient Christian is Nate Plummer, a high school biology teacher accused of teaching creationism. His tormentor is Dr. Malcolm

Trent, a retired anthropology professor and school board member, who, as the play opens, is licking his chops at the prospect of firing Mr. Plummer. It hardly counts as a spoiler to reveal that by the end of the play Dr. Trent has been thoroughly cornered by Nate's relentless logic. Filling out the cast were other school board members (one of whom is named Stegner). Initially open-minded but skeptical, they are eventually persuaded by Nate's arguments.

Let us consider a couple of excerpts, so that we may understand how these internecine philosophical debates play to the masses.

> STEGNER: Maybe we ought to get to the heart of the matter. What exactly did you teach, Mr. Plummer?
> NATE: Well for starters I just taught them about the Rule.
> STEGNER: The Rule? What Rule?
> NATE: The Rule is sort of like . . . like an exception to the scientific method. It assumes that everything in the universe is explainable in terms of natural causes. Intelligent causes are not allowed. The technical term for the Rule is "Methodological Naturalism."
> TRENT: Oh, for heaven's sake, Nate, that's not an exception to the scientific method. It's *part* of it.
> NATE: Oh, really? Our text book describes the scientific method pretty thoroughly. And yet it doesn't mention Naturalism.
> TRENT: It's implied Nate. Every scientist knows that.
> NATE: The point of the scientific method is to get factual explanations instead of ones that are based on preconceptions. The Rule is a preconception. It answers a question before it is asked.
> STEGNER: So how does this get us into a discussion of religion?
> NATE: The Rule says that we can't consider even the possibility of a designer to explain the origin of life. That kind of assumption smashes the basic beliefs of all theistic religions. (Schwabauer and Calvert 2002, 17–18)

Skipping ahead:

> TRENT: Show me the peer reviewed articles, Nate! Show me the backing for Intelligent Design Creationism in the science journals!
> NATE: But that's just the point! The science publications aren't going to print articles that challenge Darwinism. They use the Rule more stridently than anyone else.
> TRENT: How convenient! The journals are against you, eh? Some cosmic conspiracy to keep God out of Science. Even if it were true I wouldn't blame them. Just look at you! You obviously want to ram your Christian beliefs into the classroom. You and your kind do everything in your power to destroy good science.
> NATE: By explaining the Rule? All I'm doing is being honest about a hidden assumption. (ibid., 20)

Page after page of this! Normally this is where I would rail against the caricatures and distortions of ID rhetoric. Alas, in this case I am unable to do so. It has happened too often that critics of ID really have treated MN as a "rule" to which science must adhere. Robert Pennock has written:

> Put simply, the argument [presented at trial] was that as a point of method science does not countenance appeals to the supernatural. Again, we did not claim *only* science requires this ground rule. Such appeals are disallowed in court as well, for instance; MN is tacitly assumed in legal reasoning just as it is in science and should be so for the same reasons. (Pennock 2011, 546)

Philosopher Michael Ruse expresses the same sentiment, writing, "It is simply that the methodological naturalist insists that, inasmuch as one is doing science, one avoid all theological or other religious references. In particular, one denies God a role in the creation" (Ruse 2001a, 365).

These statements are perilously close to what was presented in the play. To say that science "requires this ground rule," or that "one denies God a role in the creation" while doing science, certainly makes it sound as though, no matter how strongly the evidence pointed to design, scientists, in their professional work, would simply be forced to disregard it.

At this point we might be tempted to eschew altogether any talk of demarcation. Why quibble over the boundaries between science and nonscience? Why not simply point to the gross errors in ID arguments and leave it at that? Focusing on methodological ground rules just makes life easy for the propagandists and does not address any of the concerns people have about evolution. Historian of science Thomas Dixon has written:

> In appealing to a supernatural cause as part of their science in the 21st century, ID theorists are certainly unconventional, eccentric, and out of step with recent practice, but that need not mean they are to be excluded from the realm of science altogether. There is no need for defenders of mainstream science to risk seeming ideological and doctrinaire by prejudging the kinds of entities that will feature in successful scientific theories in the future.
>
> In short, opponents of ID who use the weapon of philosophical demarcation may be shooting themselves, quite unnecessarily, in the foot. In the United States, the pro-religious intent and effect of any policy mandating the teaching of creationism or ID will be enough to keep it off the statute book. There and elsewhere, scientists and theologians, as well as voters and judges, also have many very good reasons to resist ID without straying into the fraught philosophical realm of demarcation. (Dixon 2008, 101)

As you might suspect, I am very sympathetic to this way of thinking. Unfortunately, though, demarcation arguments cannot be avoided entirely. The reason is that Dixon is mistaken about the realities of American constitutional law.

A state action can have both a religious intent and a religious effect while still being permissible.

The relevant legal standard is the Lemon test, set forth in the 1971 Supreme Court case *Lemon v. Kurtzman*.[1] The Court put forth a three-pronged test for determining if a state action violates the establishment clause of the Constitution. The first prong addresses intent and says that the action must have a secular, legislative purpose. The second prong addresses effect and says that the primary effect of the action must not be either to advance or to hinder religion.[2] An action that violates any one of the three prongs is unconstitutional. Notice that a religious purpose is acceptable so long as it is accompanied by a legitimate secular purpose. A religious effect is likewise acceptable so long as the primary effect does not involve religion.

Proponents of ID argue that their purpose is to teach the best current science and that the primary effect of doing so is to leave students better informed than they were previously. The pro-religion conclusions some might draw from ID are as irrelevant as the anti-religion conclusions some draw from evolution. In puncturing this argument it is insufficient to show simply that ID's scientific arguments are mistaken, since it is not unconstitutional to teach bad science. It is teaching religion that is unconstitutional, and part of showing that ID folks want to do precisely that is demonstrating that their ideas are not part of science at all.

Toward that end, it is reasonable to argue that ID lacks all of the characteristics and conventions of modern scientific practice. Methodological naturalism is one of those conventions. It is readily justified both by its persistent success and by the awful track record of competing investigative methods. The case against ID's scientific pretensions can thus be presented briefly as follows: ID lacks all of the attributes we expect from good science, and its arguments are far too weak to justify changing our notions of what good science entails.

I would therefore modify Dixon's argument. Demarcation talk is unavoidable in the courtroom, but that is the only place where such arguments are relevant in anti-ID advocacy.

In *McClean v. Arkansas* (1981), Michael Ruse's testimony regarding the demarcation question was influential in the judge's finding that creationism was not science. Ruse listed five criteria for distinguishing science from nonscience. For example, science relies on natural law (in contrast to the creationist's invocation of God), and science's conclusions are always tentative (in contrast to the creationist's insistence that the Bible is infallible.) After the trial several philosophers strongly criticized Ruse for, in their view, misrepresenting the relevant philosophical issues (Laudan 1982, Quinn 1984).[3] History repeated itself after the *Kitzmiller v. Dover* trial in 2005. This time it was Robert Pennock testifying for the plaintiffs, and he, too, faced the disapprobation of some of his fellow philosophers (Monton 2009).

So that there is no misunderstanding of my own view, let me say that I strongly disagree with the critics. There was nothing remotely dishonest or disreputable in the testimonies given by Ruse and Pennock. Both did a reasonable job of capturing how scientists view their enterprise and of explaining why creationism and ID fall well outside the bounds of currently acceptable practice. I am critical of some of the rhetoric that has been used in support of MN, but not of the principle itself.

My objection is to the tactically unwise move of featuring demarcation talk prominently in anti-ID argumentation. I further object to the often seen transformation of MN from a convention based on consistent usefulness into a hard and fast rule that science must respect. Philosophers Maarten Boudry, Stefaan Blancke, and Johan Braeckman have ably expressed the crucial distinction:

> In fact, there is an important divergence of opinion on the rationale of MN and its proper role in science. We will argue that the most widespread view, which conceives of MN as an intrinsic or self-imposed limitation of science, is philosophically indefensible. On that account, it is also an ill-advised strategy to counter the claims of IDC [Intelligent Design Creationism] and other forms of creationism. As we have detailed elsewhere, opponents of evolution have exploited the philosophical flaws in this popular presentation of MN to accuse scientists of philosophical prejudice and dogmatism. Alternatively, we will defend MN as a provisory attitude of science based on the unsuccessful track record of supernatural explanations. Supernatural claims do not fall beyond the reach of science; they have simply failed. (Boudry et al. 2010, 228; see also Dawes 2009, ch. 1 and Fales 2009)

This is exactly right, in my opinion.

We need to clarify what we mean by "supernatural." I would return to my example of the fish (see chapter 19). For the fish in the tank, reality is marked by certain physical constraints. For them, the walls of the tank are, so far as they are concerned, the end of the world. Their ability to alter their environment is severely limited. We humans are not similarly constrained. We can reach into their world and make radical changes, like moving the castle from one side to the other. For that matter, we can destroy their current world and create a new one for them instead. Surely the fish could reasonably describe us as "supertankular." That we are ourselves embedded within a larger natural world, constrained by all manner of physical limitations of our own, does not affect the usefulness of the term from the perspective of the fish.

In asking if the supernatural exists, I am asking if there are entities that stand in relation to humanity in the same way that we stand in relation to the fish. I am asking if there are entities that can reach into our world and make radical changes utterly beyond our abilities. If we encounter an agent with the demonstrated ability to alter fundamental constants, then it would be appropriate to describe it as supernatural. That, at any rate, would be a sufficient condition.

This, I believe, is how ID folks envision the designer they believe they have detected. If they stand accused of invoking the "supernatural," then our definition of that word must be along the lines I have described.

With this definition in hand it is clear that for scientists the relevant distinction is not natural/supernatural, it is empirical/nonempirical or testable/nontestable. I fail to see why we could not, in principle, have in mind a supernatural agent whose abilities and intentions are sufficiently delineated to produce testable hypotheses.[4] Philosopher Elliott Sober suggests as a fanciful example "the claim that an omnipotent supernatural being wanted above all that everything in nature be purple" (Sober 2008, 80).[5] I will grant that this is a silly and trivial example, but that is precisely the point. In principle you can hypothesize supernatural entities whose abilities and desires are stated with sufficient precision to permit testable hypotheses to be formed. But no one has ever managed to do that in a way that advanced scientific research. That's why it is fine to defend MN tentatively on practical grounds, but it is not fine to present it as some sort of inviolable rule.

It seems we have come full circle. We opened this discussion wondering why ID proponents are so unwilling to respect science's boundaries. Now we must wonder why some critics of ID are so emphatic about walling off science in the first place. Viewed as a convention of current scientific practice based on its long track record of success, MN seems entirely reasonable. Viewed as a fundamental ground rule to which science must always and everywhere adhere, MN seems dogmatic and unnecessary. Why, then, do so many go that route?

The answer is that protecting science from nonscience is not the only issue. Many of the most passionate defenders of MN also seek to protect religion from science. The same walls that keep pseudoscience out also serve to keep science in. Geologist Keith Miller writes:

> They [ID proponents] see methodological naturalism not as a description of the limitations of scientific inquiry, but as an arbitrary and unjustified prescription that prevents scientists from including supernatural action in their scientific explanations. However, these attacks on methodological naturalism are misguided and reveal a failure to distinguish between natural and supernatural agency, and to recognize that scientific descriptions, however complete, pose no threat to theological understanding of the action of God in nature. (Miller 2008, 118)

Robert Pennock expresses similar thoughts:

> My advice to scientists is simple: Scientists need to recognize and respect, as most do, the limits of methodological naturalism. If individual scientists wish to dive deeper into metaphysical waters, then they should be clear when they are doing so—a simple, "Now for a philosophical aside . . . ," will often do—and not suggest that their conclusions are drawn strictly from within science. If they want to keep out of trouble, however, it is probably best for scientists to leave the metaphysics to metaphysicians and theologians. (Pennock 2007, 335–336)

Both of these statements are deeply problematic.

That science is only methodologically naturalistic does not at all imply that it poses no threat to theological understandings of God's action. Consider the problem of evil, for example. The God of traditional Christian theism is said to be all-powerful, all-knowing, and all-loving. Consequently, the profligate evil and suffering in the world becomes difficult to comprehend. There are many ways of responding to this objection, but I doubt if anyone would consider it adequate to say, "The profligacy of evil and suffering is an empirical fact about the world, and such things simply have no bearing on theological truths." Consider also the question of humanity's creation. Traditional teaching holds that the world was created specifically so that human beings could live. With that as our starting point, it is rather surprising to discover we appeared as a chance end point of prolonged evolution. Clearly, there is tension here.

In each case we have a conflict between the empirical realities of nature and deeply held notions about God. Conflict arises because we are not hypothesizing incomprehensible supernatural agents, capable of doing anything, anywhere, at any time, but instead have very definite ideas about God's intentions and abilities. In saying there is a conflict, however, I am not saying there is a contradiction. The point is not that science flatly disproves the notion that God is good, or that it refutes the idea that evolution unfolded in a way that expressed His intentions. It simply makes those ideas more difficult to believe, in precisely the same way that the evidence presented in a courtroom trial makes it increasingly difficult to believe in the defendant's innocence.

That is why it is insufficient to argue that since science is only methodologically naturalistic it can pose "no threat" to theological understandings of the world. "No logical disproof," perhaps, but it can certainly pose a severe challenge nevertheless.

Which brings us to Pennock's statement. We can agree that metaphysical territory lies beyond the walls of science's fortress without agreeing that such territory is the natural domain of theologians. It is unclear what expertise they possess permitting them to speak on metaphysical questions with greater authority than what scientists bring to the table. They can speak to questions regarding their own traditions, of course, but we should reject totally the idea that they are in a privileged position for addressing metaphysical problems.

There are no experts on the ultimate nature of reality. We are all just muddling along, trying to formulate the most reasonable answers we can, based on what we know of the world. It is everyone, not just scientists, who need to be modest and circumspect when wading into metaphysical waters. It is very much in doubt, however, in comparing science with theology, that it is the former that must apologize for excessive arrogance.

21

Irreducible Complexity

Back in Knoxville, the sole afternoon speaker was the author and biochemist Michael Behe, from Lehigh University. He was halfway through his talk when we returned from lunch at the Subway sub shop (see chapter 2)—remember how long the line was? The sixteen-year-old was still lecturing me about neuroscience and brain chemistry. Since he was considerably more interesting than the formal presentation, I wound up hearing little of what Behe had to say. However, given the centrality of Behe's writing to the ID movement, it behooves us to give some consideration to his main argument.

Behe's contribution is the notion of "irreducible complexity." In *The Origin of Species* Darwin wrote, "If it could be demonstrated that any complex organ existed, which could not possibly have been formed by numerous, successive slight modifications, my theory would absolutely break down. But I can find out no such case" (Darwin 1859, ch. 6). Behe believes he has found the systems Darwin overlooked:

> By irreducibly complex I mean a single system composed of several well-matched, interacting parts that contribute to the basic function, wherein the removal of any one of the parts causes the system to effectively cease functioning. An irreducibly complex system cannot be produced directly (that is, by continuously improving the initial function, which continues to work by the same mechanism) by slight, successive modifications of a precursor system, because any precursor to an irreducibly complex system that is missing a part is by definition nonfunctional. An irreducibly complex biological system, if there is such a thing, would be a powerful challenge to Darwinian evolution. (Behe 1996, 39)

The logic of the argument seems clear enough. Natural selection only recognizes immediate utility. It lacks the foresight to preserve a currently functionless structure on the grounds that subsequent innovations will render it useful. If Behe is correct, then an irreducibly complex system serves no purpose until all of its parts are in place. This so reduces the probability of the structure having evolved by gradual selection that we can effectively dismiss the possibility. A major portion of his book provides detailed descriptions of several candidates for irreducibly complex systems. They include the vertebrate immune system and blood-clotting cascade and the flagellum of certain bacteria. He is not humble regarding the conclusion of his argument:

> The result of these cumulative efforts to investigate the cell—to investigate life at the molecular level—is a loud, clear, piercing cry of "design!" The result is so unambiguous and so significant that it must be ranked as one of the greatest achievements in the history of science. The discovery rivals those of Newton and Einstein, Lavoisier and Schrödinger, Pasteur and Darwin. (ibid., 233)

Behe's book was among the first I read upon becoming interested in evolution and creationism. At that time I knew almost nothing about blood clotting or any of Behe's other biomolecular systems. It nonetheless seemed clear to me that Behe's logic was faulty.

That *something* is fishy becomes evident from considering analogies to everyday life. Items like telephones and automobiles were luxuries when they first appeared, but are nowadays essential for most people. Their centrality to the modern world tells us nothing about the more peripheral roles they played in the past. We are also familiar with examples from architecture. There are many structures, an arch for example, that cannot support themselves until the entire edifice is in place. During construction a scaffold carries the necessary weight. When the structure is complete, the scaffold becomes redundant and is removed. The remaining load-bearing parts, previously just helpful, are now essential.

Biology provides us with numerous examples of these principles in action (Orr 1996). One such comes from a sequence of fossils documenting the transition from reptiles to mammals. The malleus and incus bones in the mammalian inner ear (more commonly known as the hammer and the anvil) bear impressive similarities to two bones in the jaw joint of reptiles. This suggests that the ear bones of modern mammals are shrunken and relocated versions of the comparable bones in the reptilian jaw joint. Such a transition seems implausible since the intermediates, lacking both ear bones and jaw bones, would be pitiable indeed, but the fossil record helpfully provides numerous specimens documenting how the transition occurred. Stephen Jay Gould explains:

> Contrary to creationist claims that such a transition cannot occur in principle because hapless in-betweens would be left without a jaw hinge, the principle of redundancy suggests an obvious solution. Modern mammals hinge their jaws between squamosal (upper jaw) and dentary (lower jaw) bones; other vertebrates between quadrate (upper jaw) and articular (lower jaw) bones destined to become the incus and malleus of the mammalian ear. Suppose that mammalian ancestors developed a dentary-squamosal joint while the old quadrate-articular connection still functioned—producing an intermediary form with a double jaw joint. The old quadrate-articular joint could then be abandoned, as its elements moved to the ear, while the jaw continued to function perfectly well with the new linkage already in place. . . .

> The abstract predictions of the last paragraph (actually advanced by paleontologists before the discoveries, so I am not just making a rhetorical point here) have been brilliantly verified in abundant fossil bone. (Gould 1993, 105–106)

By applying similar principles, biologists have uncovered plausible evolutionary scenarios for a variety of complex adaptations. These include the Krebs cycle (Melendez-Hevia et al. 1996), the vertebrate immune system (Bottaro et al. 2006), the blood clotting cascade (Miller 1999), the wings of birds (Gishlick 2004), and the bacterial flagellum (Pallen and Matzke 2006), among many others.

These ideas have been entirely standard among biologists going all the way back to Darwin. In *The Origin of Species* he wrote:

> We should be extremely cautious in concluding that an organ could not have been formed by transitional gradations of some kind. Numerous cases could be given amongst the lower animals of the same organ performing at the same time wholly distinct functions. . . . In such cases natural selection might easily specialise, if any advantage were thus gained, a part or organ, which had performed two functions, for one function alone, and thus wholly change its nature by insensible steps. Two distinct organs sometimes perform simultaneously the same function in the same individual. . . . In these cases, one of the two organs might with ease be modified and perfected so as to perform all the work by itself, being aided during the process of modification by the other organ; and then this other organ might be modified for some other quite distinct purpose, or be quite obliterated. (Darwin 1859, Ch. 6)

The relevance of these observations is clear. They illustrate the impoverished view of evolution suggested by Behe's argument, in which complex structures arise by the sequential addition of unambiguous "parts." The reality, appreciated already by Darwin, is that redundancy, followed by co-option of function, is entirely to be expected in the course of evolution. It is consequently just an error in reasoning to think that mutual interdependence in the present somehow precludes a gradual formation in the past.

Darwin was hardly the only one to notice. Writing in 1918, biologist H. J. Muller, a future Nobel laureate, wrote:

> Each new mutation in turn must have derived its survival value from the effect which it produced upon the "reaction system" that had been brought into being by the many previously formed factors in cooperation: thus a complicated machine was gradually built up whose effective working was dependent upon the interlocking action of very numerous different elementary parts or factors and *many of the characters and factors which, when new, were originally merely an asset finally became necessary because other necessary characters and factors had subsequently become changed so as to be dependent on the former.* (Muller 1918, 463–464, emphasis in original)

When you familiarize yourself with this sort of literature, you come to understand why scientists are so unimpressed with Behe's argument.

As it happens, shortly after the conference in Knoxville, I had a second opportunity to see Behe speak, this time in Washington D.C. During the question period I engaged him on these issues. I said:

> When I look at the biological literature and when I read replies to your work from other biochemists and other biologists, it seems to me that biologists who actually study these systems find a lot of evidence that they were cobbled together from simpler precursors, what Stephen Jay Gould used to call the senseless signs of history, things that make sense if you view these systems as the result of a long evolutionary process, but are hard to understand as the result of intelligent design. And I also notice that for certain specific complex systems, such as the immune system, or the blood clotting cascade, or the Krebs cycle, or the mammalian inner ear structure, or numerous others, biologists seem to have pretty good evidence for their conclusions about what the intermediate stages actually are. So my question is this: You're asking scientists to effectively abandon this idea that complex systems evolve gradually by Darwinian mechanisms. Do you have any suggestions for what they would replace it with? And I don't mean necessarily an actual fully developed theory for what they would replace it with, but just what sorts of experiments, what sorts of activities should scientists be doing in their day-to-day professional lives, that they would do under your way of looking at the world but currently do not do because they have Darwinian blinders on?

Here, in its entirety, was Behe's reply:

> Well, cobbled together is in the mind of the beholder. I'm real suspicious of somebody who says, gee that looks like it was cobbled together, when the system works pretty well, especially at the protein level. People confuse—that's why I try to separate Darwin's theory into common descent, random mutation and natural selection. Common descent—you know, one protein might resemble another protein that has a different role in the cell, but the question is not common descent, did this arise from a gene, from another protein, the question is how did this go on to have spiffy new properties. And because of the data I talked about today, I think the transition, or going from one to another is a whole lot harder than Darwinian biologists have given it credit for. So the similarities are really interesting, tracing lines of descent, that's very interesting, but inferring from that that it was a Darwinian random process, that's a whole lot trickier, and I think, like I say, people let their presumptions guide them more than they should.
>
> As to what they should do, I think for the very large part, they can probably continue to do exactly like they're doing. Because nobody's looking into

> how Darwinian processes put together complex systems, they all just presume them, they're trying to trace lines of descent, and that's an interesting thing but they're not actually trying to explain them. If I were a head of a granting agency, and if I could give money to someone I thought was doing interesting work pertinent to this area, I would give a whole lot more money to Richard Lenski,[1] or other folks who do similar types of experiments, to try to answer the question in finer and finer detail exactly what can random processes do. Let's stoke up Lenski's work, he worked with 10mL flasks, let's give him a 100 L fermenter, let's give Michigan State a multi-investigator grant so that when Lenski retires or goes to Florida, someone else can watch this over longer and longer periods of time. Because in my view the more we know about, the more we watch actual evolution in nature without models that govern our presuppositions, the more and more I think we will see that Darwinian processes are, in fact, limited.

There is much to object to here. For example, my use of the phrase "cobbled together" was not a judgment on the functionality of the systems in question. That they work "pretty well" is not the point. I was simply noting that the universal observation of scientists is that complex adaptations are never pristine creations from nothing, but are instead made from alterations to parts readily available in the organism's closest evolutionary cousins. They have just the structure necessary for natural selection to be a viable hypothesis. Seems a bit coincidental if they are actually the result of intelligent design.

Of greater significance, however, is Behe's admission that little, if anything, would change in daily scientific practice upon adopting his view. What happened to ID being a new scientific paradigm, one that will liberate science from the dead end of Darwinism? Behe is welcome to demur from the general consensus, but if his ideas have no relevance to daily practice it is unclear why scientists should notice.[2]

In the interest of wrapping this up on a lighter note, let me mention something that happened shortly after the events of the previous section. I was participating in a chess tournament in southern Virginia. During one of the nights of the tournament, the cable network C-Span broadcast Behe's presentation, including the Q&A. The following day, while a few of us were milling around in anticipation of the next game, one of the other players came bounding up to me. "I saw *that* guy on television last night!" he said, pointing at me and smiling broadly. Speaking directly to me he added, "You were giving it to that creationist guy. And I notice that he never really answered your question, did he?"

PART IV

Visiting the Creation Museum, Petersburg, Kentucky, August 2010

22

Creation Cinema

A museum employee stood at the front of the theater and said:

> This show is a little different than the rest of the museum. This is a satire, meant to lighten us up a little, but still teach us some very important facts. You might recognize some of these stereotypes from your school days. So relax and enjoy this unique experience. We will now join Wendy at her campfire, as she ponders some of life's most perplexing questions. Prepare for some fun and some surprises. Prepare for some solid answers. Prepare to believe.

Wendy, it turned out, was an animatronic teenager sitting aside a model campfire. Her head moved slowly up and down as the theater lights dimmed. A nighttime scene was projected in front of her.

"Does anybody even know I'm here?" she asks. The smoke from the campfire forms a question mark on the screen.

"Is there any meaning?" This time the smoke forms the question, "Why?"

"Did God create all this or did we just invent God?" "God" appears in smoke font.

Two small lights, reminiscent of Tinkerbell, fly in. They become two angels, Michael and Gabriel. After getting Wendy's attention by knocking on their side of the screen, Michael says, "You know you didn't really invent God. Kind of absurd when you really think about it. God actually invented you, and everything else for that matter. Everything—the whole enchilada!"

But Wendy is still mired in existential quandaries. "Of course I believe in science. Science is how we cured polio and invented cell phones." The angels interject to explain that God loves science and even exhorts us to explore the world He created. Michael questions whether Wendy is listening.

She says, "Life is probably just a big accident. A predictable result of an infinite number of matter/antimatter asymmetric collisions."

Gabriel says, "Oh, she's listening, Mike. Just not to you. Problem is she can't stop listening to the voices of her culture."

The yearbook for "Enlightenment High School" appears. It opens to a random page. The promised stereotypes enter the story. . . .

I was in the "special effects theater" at the Creation Museum in Petersburg, Kentucky, watching the film *Men in White*. The museum, designed and built by the Answers in Genesis organization we met at the Creation Mega-Conference, opened in the summer of 2007, with an initial price tag of twenty-seven million dollars. Just under three years later, it celebrated its three-millionth visitor. My first visit occurred shortly after it opened, and I have been back several times since. The museum has much to teach us about modern creationism. It is also surprisingly thought-provoking, if not quite in the way the directors intend.

In *Men in White,* the yearbook opens to a photograph of Mr. P. Snodgrass, a science teacher dressed in a white lab coat. He says, "Pay attention, Wendy. This is the most important concept in modern biology. It's called evolution." His voice was a passionless monotone. Another science teacher appears, a shrieking schoolmarm named Miss E. Certainty. She is going on about "billions of years" and "natural processes." Dr. J. Plumsure, a biology professor with a pipe and a tweed coat, joins them. Soon all three are rambling as their images swirl around the screen, and their voices echo ominously. At one point we get a close-up of Professor Plumsure, who says, "There is no God who had anything to do with it!"

One of the yearbook photos, shown in silhouette and representing a random student, says, "And what if I don't believe your ideas?"

"Well, then you're in violation of the Constitution of the United States and the separation of church and state!" shrieks Miss Certainty. Her voice is shrill to the point of physical unpleasantness.

The yearbook slams shut. The angels describe this all as a tragedy, telling us bluntly that evolution leads to despair. They say, "Just think how it would change her life if Wendy found out there really is purpose and meaning to her existence!" The scene changes to that of an ersatz movie set, with Gabriel directing.

Cue the six days of creation. God places humanity in the Garden of Eden, where everything is sweetness and light and there is no death or disease. Gabriel says, "Man was without sin and could walk right there with God." After a dramatic pause he adds, "Until the rebellion."

Then I learned why they called it a special effects theater.

After a montage of snakes and lightning and sounds of biting into an apple, we learn that humanity's corruption prompted God to destroy the Earth. We see pouring rain and violent ocean waters, while Gabriel and Michael, now wearing raincoats, read portions of the Noah story. Some might have considered this sufficiently compelling, but not our hosts. To really make you feel like you are right there experiencing the flood, a motor inside your seat vibrates unpleasantly, and a small jet of water hits you in the face, fired from the back of the seat in front of you.

This was met with squeals of surprise and sporadic laughter. My own feelings were ably expressed by the four-year-old a few seats to my left. With a tone of surprise and dismay, she said, "I just got wet, Mommy."

Next up was straightforward creation science. Gabriel says, "If you use the Bible as your starting point, Wendy, then everything makes sense! But if you buy into molecules-to-man evolution—"

Michael interrupts, "Goo to you we like to say."

Unimpressed, Gabriel continues, "Uh, yeah, this goo-to-you idea of evolution, no matter what you see in the real world, you have to interpret the data through one basic foundational lens."

While I was ruminating about everything that was wrong with that, the scene changed to a conversation between television host Suzee Teevee and Dr. Ed U. Kaded. The professor gushes about the marvels of radiometric dating. Wendy is impressed, but Gabriel sets her straight: "That's a dating method based on a ton of *assumptions*. New studies are suggesting radically younger ages for those rocks."

Then Dr. Kaded says, "And there really can be no questioning of Charles Darwin anymore. I mean, virtually every thinking person on the planet agrees wholeheartedly with everything he ever said! But unfortunately knowledge itself is attacked when those 'creationists' try to get into our schools and question Darwin's conclusions."

Ms. Teevee fawns. "Isn't it unbelievable! I mean, where would we be without knowledge?"

Wendy is impressed again. "I guess if you question evolution then you must be an idiot."

Gabriel is having none of it. "What's idiotic is not to question!" Then it is back to the bullet points, this time involving something about helium diffusion in rocks and more dumping on radiometric dating.[1]

Gabriel says, "Of course, there is one place that you must never, ever, bring up any of these observations, any of this science, any of these questions."

"Where's that, Gabe?" asks Michael.

No doubt you can answer Michael's question for yourself.

School.

The angels are now sitting in the back of Mr. Snodgrass's class. They look extremely bored as they listen to a lecture about the Earth's antiquity. Seeking to liven things up, they challenge each of their teacher's points. With each challenge Mr. Snodgrass becomes increasingly impatient and dogmatic. The angels remain calm and reasonable throughout. Here is a typical exchange:

Gabriel says, "Did you know that a number of Ph.D. geologists are reconsidering how old the Earth might be? There's a lot of observational evidence out there that has convinced them that the Earth might in fact be quite young."

"Well," huffs Mr. Snodgrass, "I'd say forget about those geologists then. They must not be serious."

"Not serious because they don't agree with you?" asks Michael.

The angels make frequent use of words like "cool," "whoa," "awesome," "gnarly," and "dude," thereby leaving no doubt regarding the locus of hipness in *this* conversation. Eventually Mr. Snodgrass is reduced to the inevitable angry sputtering, "There is no God who intervenes in the world!"

Such was my introduction to the Creation Museum. As it happens, this was just one of several films I would see. While touring the exhibits it seems like every few minutes you are placed in front of a screen, passively absorbing the latest in creationist rhetoric.

There was also a planetarium show. Having visited the Hayden Planetarium in New York City, I was unimpressed, but the children on hand seemed more enthusiastic. The show itself was mostly a presentation of standard astronomical facts ("Betelgeuse is a red supergiant, roughly six hundred times the diameter of the Sun. . . . "), coupled with frequent insertions of God-talk ("Galaxies come in a variety of shapes and sizes. Each one is a glorious demonstration of God's limitless power."). There were snide references to what "secular astronomers" think. Occasionally some banal observation was tossed off as representing a fundamental problem for modern astronomy. I will spare you both the details and their inevitable refutation.

During one of my visits, the two people sitting next to me were an older, married couple. When the show ended, the woman looked at me and asked what I thought of the presentation. Not being in the mood for an argument, I replied simply that I found it interesting. "How about you?" I asked.

"I don't know," she said with a shrug. "The graphics were nice, but I could have lived without all the God-talk." She looked me dead in the eye.

Certainly not what I expected. I smiled and said, "I know just what you mean." We chatted for a minute or two. She told me she was visiting her niece, who lived nearby. Said niece, more religious than her aunt, recommended a visit to the museum. It quickly became clear that I was hardly the only ringer there that day.

23

Creation and Corruption

Have no illusions about the level of professionalism involved at the Creation Museum. This is not a fly-by-night operation or an amateurish set-up run out of someone's basement. You would never mistake it for the Smithsonian, but the exhibits look good and the environment is generally pleasant. Visiting is a full-day affair. If you arrive when it opens and stay until closing, you will probably see everything. Otherwise you might want to plan on two days. The museum grounds feature an attractive botanical garden, complete with walking trails and a petting zoo.

Much of the museum is directed toward children, and, on each of my visits, there were numerous private school groups in attendance. For someone of my general perspective, this is rather disturbing. Though I am certain the museum's directors would disagree, it is fair to say that many of the exhibits demonize science and scientists. There is a line that is crossed when the desire to instruct your children leads to hostile and dishonest characterizations of large groups of people.

While many of the exhibits relate to science generally and evolution specifically, the museum is primarily devoted to a particular view of the Bible. That leads to the philosophical question of whether "museum" is the proper term. The appeal of a real museum is the opportunity to see actual physical exhibits and artifacts. By viewing such things, you experience something you cannot get just from reading a book. Walk among actual fossils, say, or bits of memorabilia from Charles Darwin, and you feel connected, if just in a small way, to history. There is very little that is comparable at the Creation Museum. It is as though you have two equivalent options. You can stay home and read creationist literature, or you can go to the museum and see it in placard from.[1]

On each of my visits I have engaged in conversations with other attendees, as always never making a secret of my views. Following the pattern set by my conference experiences, I was never made to feel uncomfortable or unwelcome as a result.

FIGURE 23.1 The creation Museum grounds.

Others have reported less pleasant experiences, however. In June 2008 sociologist Bernadette Barton visited the museum with fourteen of her students. One of them later reported the following story:

> Hank and I started our walk through together, but not even five minutes into it we noticed that we were being glared at by everyone because I am a small, blond, white girl and Hank is a good looking black man, and for whatever reason it seemed as though the two of us couldn't believe the same thing everyone else in the building did. So Hank and I parted ways and decided to finish the walk through alone. While I was coming up on the Noah's Ark part of the story the rent-a-cop and his pet decided to walk around me twice, making it a point for me to know that I looked out of place, which made me uncomfortable even more. As soon as the pair walked away from me a man with his two kids walked up to me and said, "The reason he did that is because of the way you're dressed. We know you're not religious, you just don't fit in." I was wearing a pair of leggings and a tan long sleeve shirt that came down well past my behind. I saw nothing about the way I dressed to be out of the ordinary but I guess it was. After that honestly I couldn't have got out of there fast enough. (Barton, forthcoming)

Most of Barton's other students, some of whom were homosexual, had similar stories to tell.

* * *

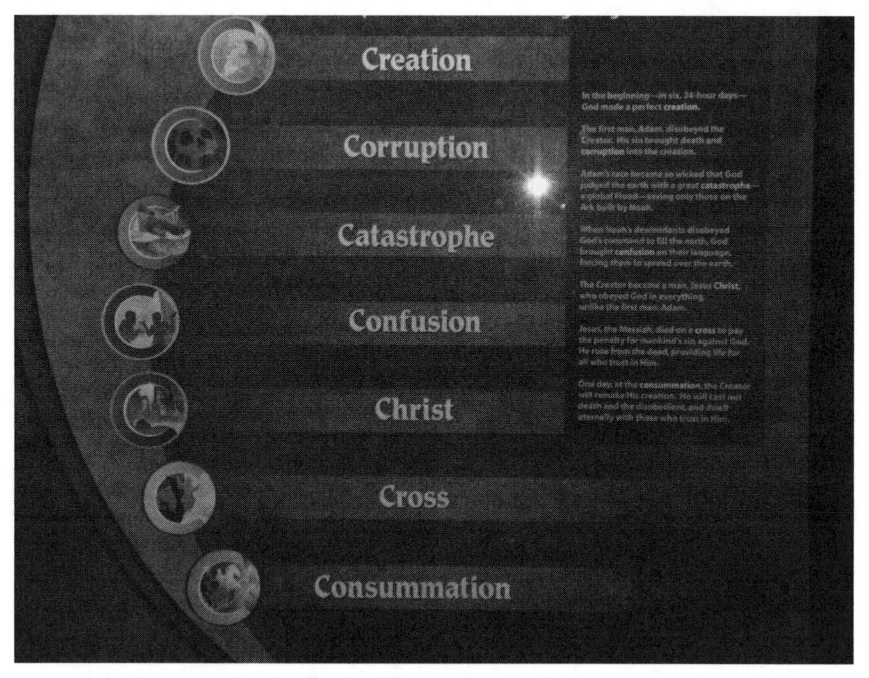

FIGURE 23.2 As presented at the Creation Museum, the most important events in world history can be summarized in these "Seven Cs."

The usher scanned my ticket and let me through. My tour of creation had begun.

I came to a diorama showing two gentlemen digging for fossils in a sandy dig site. Both were sitting on the ground. One was a balding Caucasian gentleman, roughly fifty. He was sitting ramrod straight near the center of the exhibit. Off to one side was his Chinese friend Kim, slumped over and looking a bit hangdog. On a monitor overhead the Caucasian fellow tells us that while he and Kim are both scientists pondering the same facts and evidence, they approach their work from different starting points. He is a creationist, while Kim is an evolutionist. Facts do not come preinterpreted, you see, but must be sifted through our various worldviews.

That theme persisted through the following room. A series of displays proclaimed that the facts are all in the present. We must draw inferences regarding their evidential significance for unraveling the past. A series of posters contrasted creationist and evolutionist interpretations of various bits of scientific data. (For example, scientists see fossils as evidence for evolution, while creationists see evidence for Noah's flood.) This is nothing we have not seen already save for a novel twist. The creationist view was said to rest on a foundation of "God's Word." The foundation for the evolutionist view? "Human Reason." This surprised me, since it implied that human reason is in some way hostile to creationism.

FIGURE 23.3 Adam and Eve.

Soon we come to a wall of actual magazine and newspaper headlines illustrating the pernicious effects of compromising scripture. "Scripture abandoned in the culture," the header intones, "leads to relative morality, hopelessness and meaninglessness." The headlines are about pornography, gay marriage, evolution, euthanasia, stem cell research, sex education, lack of belief in an afterlife, and school shootings.

Through the broken window of a church, we see a family listening to a compromising preacher. We are then shown videos of the same family at home. The teenage girl is seen on the phone discussing the possibility of getting an abortion, while the boys are doing drugs and looking at pornography. Lest you miss the point, a sign proclaims, "Scripture compromised in the church, leads to scripture abandoned in the home."

After a short film based on a dramatic reading of Genesis 1, I was standing in the Garden of Eden! Adam was there, dutifully naming the happy animals. Eve appears, splashing around with Adam. Both are happily naked, with Eve's long hair addressing any concerns you may have about modesty. Eve is looking down, while Adam, standing straight, is caressing her cheek.

Sadly, the happiness and bliss did not last long. Adam and Eve disobeyed God. Sin entered the world, and with it death, suffering, and pain.

24

Groaning under the Curse

Pride of place among theological problems must surely go to the problem of evil. That there is something incongruous in the picture of a just and loving God presiding over a world of extravagant cruelty and suffering is obvious to even the most unreflective person. Philosophers have given the problem precise, technical formulations (Drange 1998, Howard-Snyder 1996, Rowe 2001), but it is doubtful that David Hume's blunt version has ever been bettered for clarity or force. He writes: "Is he [God] willing to prevent evil, but unable? then is he impotent. Is he able, but not willing? then is he malevolent. Is he both able and willing? whence then is evil?" (Hume 1779, part X).

Creationists offer a clear answer to this problem. Evil and suffering, they argue, are consequences of human sin. They were not part of God's very good creation. Such good and beauty as remain are glimpses of how God meant for us to live. The more unpleasant commonplaces of daily existence represent God's tough love, as he forces us to confront, in a measure far less severe than we deserve, the results of our choices.

To Hume they might reply: "God is able but not willing to eliminate suffering, but that does not reflect poorly on His goodness. Quite the contrary. Just as it is an act of love for parents to discipline their child, so, too, is it an act of love for God to give us fair warning of the consequences of our disobedience. That He offers us the free and undeserved gift of salvation and eternal life is all the evidence one needs of His extraordinary goodness and mercy."

As evidence of their seriousness, consider the following remarks from Ken Ham. Ham's brother, Rob, died as a young man from a rare brain disease. Ham writes:

> Here was my brother, one of God's faithful children, afflicted with a disease most of us don't even want to think about—a disease that caused him to lose his mind and die slowly, while others lived on in health. My struggles echoed those of David, who in Psalm 73:12–14 said, "Behold, these are the wicked; and always at ease, they have increased in wealth. Surely in vain I have kept my heart pure, and washed my hands in innocence; for I have been stricken all day long, and chastened every morning."

> Fueled by frustration and feelings, these thoughts were tossed around in my mind. But as I continued to stand back and see the big picture (putting our specific situation into the broader biblical perspective), I was led to a conclusion that reflected God's perspective rather than my human one. Did Rob deserve to suffer the way he did? The answer is "yes."
>
> When you think about it from a Christian perspective, we *all* deserve much, much more than the suffering afflicting Rob. Because of our rebellious condition, we don't even deserve to live. But God didn't annihilate us. He has allowed us to live—while at the same time giving us a taste of what life is like without God. (Ham 2007, 90–91)

That's hardcore. Given the central role creationists give to human sinfulness in understanding the fallen state of the world, you can understand why they would be hostile to any scheme that places millions of years of suffering and death prior to the arrival of humanity.

Christians who accept evolution will be unable to believe the creationist response to the problem of evil. They will therefore need to find a solution that is consistent with the findings of modern science. Philosophers and theologians have offered a variety of solutions in this regard, and we shall consider some of them in the remainder of this chapter.

We should begin by noting that to fully understand the problem of evil, we must begin with two fundamental distinctions. The first is between the "logical" and the "evidential" forms of the argument. The former is the claim that the statements "God exists" and "Evil exists" are contradictory. They cannot both be true. On this view, granting the existence of evil entails rejecting the existence of God.[1] In this context, "God" refers to an omnipotent, omnibenevolent, and omniscient being who, among other things, created the world. The God of Christian theism is commonly said to have all of these attributes. In principle we could defuse the problem by rejecting any of the three *omni*'s, but many would consider that too high a price to pay.

The logical form of the problem of evil has had many defenders over the years, but the consensus of modern philosophers is that it is overly ambitious.[2] The basic difficulty is easily illustrated by analogy. A parent might allow a young child to suffer temporarily—say, by forcing him to undergo a painful medical procedure. The child likely does not understand what is happening, but the parent, wiser than he, sees the big picture. It is possible that from the perspective of omnipotence our sufferings are like those of the child. They are endured for reasons beyond our comprehension, but ultimately serve some greater good. Defenders of the logical problem of evil must show this to be impossible, but success in that project has proven elusive.[3]

Attention has therefore turned to the evidential argument, which holds simply that horrendous evils are strong evidence against the existence of God. By weakening the conclusion, the argument increases greatly in force. It is no longer necessary to establish the logical impossibility of God's having a morally compelling reason for permitting evil, but only that it is implausible that He has one.

The second fundamental distinction is between "moral evil," by which we mean the bad things people do to one another, and "natural evil," which refers to the harm caused by nature in full dudgeon. I am using the term "evil" a bit loosely, as a catchall phrase for anything that makes us think, "Why would God allow *that*?" It might be technically more correct to distinguish "evil" from "suffering," since the former implies an element of moral judgment while the latter does not. I do not think my usage will cause confusion, however. It seems reasonable to treat a tsunami that kills a hundred thousand people as an example of natural evil, even though it makes no sense to describe the tsunami itself as evil.

The argument that evil is a consequence of human sin is hardly original to creationists. It is the traditional Christian reply, going back to Augustine. Since evolution implies eons of death and suffering prior to the appearance of humans, accepting evolution entails rejecting this explanation.[4] Philosopher Philip Kitcher writes:

> Many people have been troubled by human suffering, and that of other sentient creatures, and have wondered how those pains are compatible with the designs of an all-powerful and loving God. Darwin's account of the history of life greatly enlarges the scale on which suffering takes place. Through millions of years, billions of animals experience vast amounts of pain, supposedly so that, after an enormous number of extinctions of entire species, on the tip of one twig of the evolutionary tree, there may emerge a species with the special properties that make us able to worship the Creator. . . . Moreover, animal suffering isn't incidental to the unfolding of life, but integral to it. . . . Our conception of a providential Creator must suppose that He has constructed a shaggy-dog story, a history of life that consists of a three-billion year curtain-raiser to the main event, in which millions of sentient beings suffer, often acutely, and that suffering is not a by-product but constitutive of the script the Creator has chosen to write. . . . Indeed, if we imagine a human observer presiding over a miniaturized version of the whole show, peering down on his "creation," it is extremely hard to equip the face with a kindly expression. (Kitcher 2007, 123–124)

Philosopher David Hull is even more blunt:

> Whatever the God implied by evolution and the data of natural history may be like, He is not the Protestant God of waste not, want not. He is also not a loving God who cares about His productions. He is not even the awful God portrayed in the book of Job. The God of the Galapagos is careless, wasteful,

indifferent, almost diabolical. He is certainly not the sort of God to whom anyone would be inclined to pray. (Hull 1991)

Natural evil was a theological problem long before Darwin, of course, but evolution brings into focus the inadequacy of any reply based entirely on human concerns. For example, Richard Swinburne offers a complex argument justifying natural evil on the grounds that it is necessary if human beings are to learn certain facts of moral significance (Swinburne 1978). John Hick has argued that natural evil is necessary for "soul-making," which is roughly the idea that we grow spiritually and morally by showing courage in the face of great suffering and evil (Hick 1966). Swinburne and Hick have been strongly challenged by Parsons (1989) and Kane (1975), respectively. More relevant to our present concerns, however, is that their arguments need considerably more developing before they can account for the suffering inherent in the evolutionary process.

One approach to this problem, favored by many philosophers, is that of "theistic skepticism (Gale 2007)."[5] We make no attempt to explain why God permits evil, but argue instead that our finite human perspective is inadequate for comprehending such things. My cat is incapable in principle of understanding why I inflict upon her the suffering and misery of a yearly trip to the vet. But I do it anyway, because my understanding of her situation is far greater than her own.

This is a viable option for our present situation, but there is certainly much to be said against it. For one thing, it essentially concedes that we do not, in fact, have a plausible way of reconciling evil with God's goodness. It also leaves us in danger of becoming complete moral skeptics. If horrific evils find their justifications in greater goods known only to God, then what confidence can we have in any of our moral judgments (Russell and Wykstra 1988)? It is sometimes said that theology is "faith seeking understanding." Abandoning that quest in the face of a fundamental challenge to theism seems like a serious abdication.[6]

Alternatively, we could try to provide a "theodicy," which is to say that we attempt to explain why God permits horrific evil. The classic example of a theodicy is the free-will defense, in which we argue that the possibility of humans committing terrible crimes is the price we pay for the greater good of free will. There have been many suggestions for evolutionary theodicies, all of which deserve more detailed treatment than I can provide here. In considering a few possibilities, however, we can at least suggest why this remains a sore point for Christians struggling with evolution.

We might hypothesize that God was logically compelled to do His creating by Darwinian mechanisms. Biologist Richard Dawkins (1983) has argued that natural selection is the only mechanism capable in principle of producing complex adaptations. If this is true, then to the extent that God wanted to create complex, well-adapted creatures it might be that he was forced to do so via Darwinism. Philosopher Michael Ruse writes:

> The Christian positively welcomes Dawkins's understanding of Darwinism. Physical evil exists, and Darwinism explains why God had no choice but to allow it to occur. He wanted to produce designlike effects—without producing these He would not have organisms, including humankind—and natural selection is the only option open. Natural selection has costs—physical pain—but these are costs that must be paid. (Ruse 2001b, 137; see also Peacocke 2001, 36–37)

This argument, however, is incomplete as it stands, since evolution by natural selection is plainly not the only option for creating human beings. God might have created everything directly and supernaturally, precisely as the Bible says He did. We must ask, therefore, what divine purpose could not have been fulfilled by creating in such a manner. However you imagine God's creative activity, it is clear that He did some things supernaturally and allowed other things to unfold by natural laws. The only question is the balance He employed. It is unclear why a world in which God restricted his creative activity to establishing the initial conditions at the Big Bang is any more lawlike than the one in which Adam and Eve found themselves at the end of Creation Week.[7]

We could try arguing as follows: Some things are what they are in part because of the process that created them. For example, God could not create a twenty-dollar bill that is legal currency in the United States, because only the U.S. Treasury can do that. God could, one assumes, create something that is atom-for-atom identical to a twenty-dollar bill, but that still would not be legal currency. It is possible that evolution plays a similar role in God's creation.[8] Let us imagine the physical state of the world at the moment, several billion years into the evolutionary process, when human beings first appeared. We could hypothesize that if God created the world directly in just that state, it would nonetheless lack something for not having arisen through an evolutionary process.

This, at any rate, seems to be the sort of possibility that Ruse and Peacocke have in mind. But what can we point to as the thing that would be lacking in the nonevolutionary world? The obvious candidate is the process itself. Perhaps the creation of humanlike creatures was only one of God's goals. We might suppose that He also takes pleasure in all of the other productions of the evolutionary process, taking special delight in the higher and higher "states of being" it produces.

If you find that plausible, though, I would redirect you back to the remarks of Kitcher and Hull. The processes of Darwinian evolution just are not the sort of thing in which a just and loving God can take delight. As Richard Dawkins has noted, "The total amount of suffering per year in the natural world is beyond all decent contemplation."[9] Indeed. That evolution occasionally produces something an observer might regard as beautiful is scant mitigation for the awesome quantities of suffering the process entails.

Theologian John Haught suggests a different answer:

> In fact, an evolutionary theology might agree . . . that the idea of an instantaneously complete creation is theologically unthinkable. Any universe that

> might conceivably burst into being fully formed could never have become truly differentiated from its creator. It would not have had the time or opportunity to become a world that stands out distinctly in dialogical relationship to God. Such a "world" would be a purely passive mirroring of the divine will. Indeed, it would not be a world at all, but instead an eternal dimension of God's own being. A universe sculpted to finished perfection in the first instant of its existence would be frozen in place forever. It could not give rise to beings endowed with freedom or even with life, since by definition living and freedom-endowed organisms are inherently self-transcending realities whose very nature is to move beyond their present state of being. (Haught 2008, 174)

The force and confidence of Haught's prose may blind us to the fact that what he describes as theologically unthinkable was essentially what most theologians thought for a great many centuries. One wonders how so many Christian scholars overlooked the unthinkability of such a central notion, that of a primordially perfect creation sullied by human sin.[10]

At any rate, it is simply false to say that a universe created in a state of initial perfection is then frozen in place. In the traditional view, God's creation was initially perfect, but was then placed in the care of sinful human beings. God gifted us with freedom, thereby ensuring that the world would not long remain a passive mirroring of the divine will. It quickly became instead a reflection of human will, much to its detriment. It was human freedom, not billions of years of cosmic and biological evolution, that permitted the world to become distinct from its creator. Haught's statement implies that human freedom itself is something that can only arise through evolution, but it is hard to imagine why that would be the case. Why is it impossible for a human being created directly by God to be an "inherently self-transcending reality" whose nature moves him beyond his "present state of being"?

Central to this argument is the idea that the sorts of natural laws leading to Darwinian evolution are the only ones that could have fulfilled God's purposes. But what reason is there for thinking that? Even granting that it suited God to do His creating through a lengthy, law-bound process, why did that process have to be specifically Darwinian?

It is difficult to speculate about alternative systems of natural laws, since none of us has any experience creating life-sustaining universes. Let us agree, then, not to be too dogmatic one way or the other. My own view is that the theologians and philosophers have not been successful at all in establishing what they need. In my view Philip Kitcher has it exactly right:

> When you consider the millions of years in which sentient creatures have suffered, the uncounted number of extended and agonizing deaths, it simply rings hollow to suppose that all this is needed so that, at the very tail end of history, our species can manifest the allegedly transcendent good of free and virtuous action. There is every reason to think that alternative processes for unfolding

> the history of life could have eliminated much of the agony, that the goal could have been achieved without so long and bloody a prelude. (Kitcher 2007, 127)

Sadly, an expression of disbelief, no matter how cogent, is no substitute for an argument. That is why the professional literature now records a secondary debate about who bears the burden of proof. Do atheists have to provide an alternative system of natural laws that is demonstrably superior to what we have? Or are theists required to show that no such alternative exists? Writing in 1930, theologian F. R. Tennant placed the burden on the skeptic: "That painful events occur in the causal chain is a fact; but, that there could be a determinate evolutionary world of unalloyed comfort, yet adapted by its law-abidingness to the development of rationality and morality, is a proposition the burden of proving which must be allotted to the opponent of theism" (Tennant 1968, 201). Bruce Reichenbach (1982) also defends this view. But McLean (1991), though defending theism against the problem of natural evil, nonetheless argues that the theist does need to assume the burden of proof. Trakakis (2005) argues that it is unreasonable to expect atheists to describe a functional universe from scratch, but that strong reasons can be given nonetheless for thinking that our system of natural laws is not ideal. Perkins (1983) argues along similar lines.

Philosopher Luke Gelinas aptly summarizes the situation:

> The question, then, is whether we are justified to conclude that the worlds the critic imagines—those very much like our own, minus the laws responsible for natural evils—are truly possible. There is a tendency at this point for participants in the debate to become dialectically mired in (what Richard Gale dubbed) a Modal Intuition Bowl, with the critic of the natural law theodicy insisting that such worlds are possible, the defender of the strategy vehemently denying as much. (Gelinas 2009a, 546)

Well said. We cannot show conclusively that the natural law theodicy is false. That acknowledged, I certainly know where my own modal intuition lies.

Theologian Christopher Southgate, while endorsing the general principle of Ruse's argument (that creation via Darwinian evolution was logically necessary for achieving God's aims), notes a further difficulty. He directs our attention to those creatures whose life is "all suffering and no richness" (Southgate 2008, 40). It is morally insufficient to tell the creature that its suffering was, at least, a link in an evolutionary chain. Referring to previous work by physicist R. J. Russell, and describing Ruse's claim as the "only way" argument, he writes: "My own view is that we must still presume a version of the "only way" argument. . . . But I agree completely with Russell that this move by itself is not enough, since it does not answer in any way the fact of suffering at the level of the individual creature" (ibid. 48).

His solution is to propose "a compound evolutionary theodicy" of which the "only way" argument is just one part. It is supplemented with the idea that God suffers with His creation, with this suffering reaching its highest form in the Cross of Christ. He then imagines that the suffering of animals in this life is redeemed in the afterlife, an idea he terms the "pelican heaven" argument (ibid. 16). Animals that know no flourishing in this world can find it in the next.

The question of whether God can reasonably be said to suffer along with His creation is difficult. Can an omnipotent being suffer? The best analogy I can devise once again involves my cat. When I am driving her to the veterinarian, it is heart-breaking to hear her forlorn and defeated meows coming from the back seat. It is immensely frustrating and distressing to me that I cannot explain to her what I am doing. In this sense I can, perhaps, be said to suffer with her. But I also find a certain satisfaction in accepting this sort of suffering. Specifically, it is the satisfaction of knowing that I am doing the right thing for her even though it is difficult. I also know that her very temporary suffering at the vet is compensated for by the flourishing that is made possible by keeping her in good health.

The notion of "pelican heaven" does have theological problems, though. Animals are typically thought to lack souls, making it unclear what part of them actually survives their bodily death. There is also the problem that some animals flourish in part by inflicting misery on other animals. Does lion heaven come equipped with antelopes to slaughter? We might also wonder why God bothered with the physical world at all. Why not just create heaven in the first place?

Moreover, it is very debatable whether hypothesizing an animal afterlife really solves the problem. Can the concept of "flourishing" be applied to animals? Even when things are going well, their lives are usually defined by an endless search for food and mates and by a perpetual fear of predators. Does grimly tending to the necessities of their existence constitute "flourishing" for them? If we imagine a rabbit afterlife, say, in which the creatures have no fear of predators or starvation, are we to suppose that living in such luxury is adequate recompense for an earthly life that was all suffering and no richness?

Humans can create music and art, they can contemplate weighty questions, and they can leave the world much different from how they found it (for better or for worse). They can even understand how times of suffering and adversity can lead to greater goods later on. Animals, so far as we know, cannot do such things. They merely exist. They can experience the lows of pain and suffering, but it is unclear if they are capable of experiencing sufficiently compensatory highs.

Southgate addresses some of these objections (ibid. 78–91), and I will refer you to his book for further discussion. For me the really salient point lies in a different direction. Our thinking has become hopelessly abstract. When we were pondering the "only way" argument, at least we were still somewhat tethered to the real world. The question of whether God had options other than Darwinism is informed by science, even if, in the end, it cannot be definitively resolved scientifically. But once we start hypothesizing pelican heaven into existence, we have simply left behind any connection with empirical realities. I have no doubt that a sufficiently clever theologian, arguing from his armchair, can summon forth answers to any objections I might raise. The difficulty comes when you then try to persuade yourself of the correctness of his speculations.[11]

To me it seems that theological speculations about God being forced to create through Darwinian mechanisms, or hypotheses about pelican heavens, arise because they are necessary for preserving the assumption of God's goodness, not because there is a compelling argument to be made on their behalf. If you are possessed of sufficient imagination to find such things plausible, then you are welcome to them. I have to wonder, though, if these are really the sorts of arguments creationists are deemed theologically naive for rejecting.

Maybe we are looking at things the wrong way. Our attitude has been that modern theories of evolution must somehow be reconciled with traditional conceptions of God. From where do those traditional conceptions come? They are certainly not straightforward inferences from the Bible, since any text gets interpreted through the lenses of particular cultures and biases. Early Christianity very quickly migrated to Greece, where nascent ideas about God and Jesus suddenly encountered a particular strain of philosophical thought. We should consider the possibility that much of what we regard as bedrock Christian doctrine actually represents the commingling of the biblical witness with Greek philosophy.

In this regard, theologian Robert Mellert writes, in reference to different schools of Greek philosophy:

> By a fateful choice of history, Parmenides became the father of metaphysics and the basis for later Greek philosophy, while Heraclitus was largely ignored. As a result, the thrust of Greek thought, and most of Western thought thereafter, was derived from the static concepts of "being," "substance," and "essence," rather than the more dynamic concepts of "becoming," "process," and "evolution." (Mellert 1975, 12)

This influence is felt in the traditional conception of God as static, unchanging, and ultimately aloof from the world. That has always been theologically problematic. For many Christians, a central notion is that of having a "relationship" with God. Now, when we speak of a relationship between two people, or even between a person and a beloved pet, it is implied that both parties are changed in some way as a result. How, then, could we have a relationship with a perfect, unchanging God? How could He possibly be affected by anything humanity does?

Evolution, far from being a threat to Christianity, can thus be seen as a badly needed corrective to the errors in past thought. It focuses us on the centrality of change, on the importance of "becoming" over "being." God does not create the world in a state of perfection and then watch it wind down over time. Instead He sets in motion a process that leads to higher and higher states of being. He does not push us away from the past, He draws us toward Him from the future.

This sort of thinking lies at the heart of "process theology," a movement that achieved a measure of prominence among academic theologians in the 1960s and continues to have adherents today (Plantinga et al. 2010, 546–548). It forces a reconsideration not just of doctrines surrounding creation, but of even more fundamental considerations regarding divine omnipotence and the nature of Jesus.[12] In particular, process theologians generally reject the notion of Jesus as "the God-man." They argue that He was fully human, but was far more open than most to the possibilities made available to humanity by a loving God (Mellert 1975, 75–88).

John Haught now explains the importance of this sort of thinking:

> The fact of evolution, however, introduces a dimension that previous responses to the problem [of evil] did not have available to them. This is the simple fact that the universe is still in the process of being created. It is not yet finished, and if it is not yet finished it cannot yet be perfect. An evolving universe may aim toward perfection, but at any moment prior to such an unimaginable fulfillment it will have to be not-yet perfect. And if it is not perfect, then we cannot be altogether surprised that imperfection, including the fact of pain, will be a part of it. (Haught 2000, 55–56)

Of course, those of a more conservative temperament will simply find process theology entirely unacceptable. They will not simply reject the idea that traditional teachings about God and Jesus are contaminated by Greek philosophy, but will do so with an expression of withering contempt.

Also problematic is that Haught's view is based on the very controversial notion that evolution is progressive. He writes,

> Without too much difficulty, we can make out a kind of story line along which nature has traveled from trivial to more intricate and eventually sentient, conscious, and self-conscious states of being. . . . It is hard even for the most entrenched pessimist to discount altogether the obvious "directionality" visible in the overall movement of the cosmos from simplicity to complexity. (ibid. 117)

Few biologists and philosophers would agree that directionality in evolution is obvious. In fact, the general consensus is against this view.[13]

The most serious objection, however, is simply that Haught's argument seems nonresponsive to the problem. The savagery of the evolutionary process cannot be seen as a stepping stone from a chaotic past to a glorious future. We are not evolving our way out of it. It is entirely possible that a few thousand years from now humanity will be extinct, but whatever species remain will still be involved in a struggle for existence. Natural selection will continue to reward them for whatever bits of sadism and malice facilitate the insertion of their DNA into future generations.

An especially bold attempt to address this problem comes from biologist Francisco Ayala. He argues that, far from exacerbating the problem of evil, evolution actually helps defuse it:

> Indeed, a major burden was removed from the shoulders of believers when convincing evidence was advanced that the design of organisms need not be attributed to the immediate agency of the Creator, but rather is an outcome of natural processes. If we claim that organisms and their parts have been specifically designed by God, we have to account for the incompetent design of the human jaw, the narrowness of the birth canal, and our poorly designed backbone, less than fittingly suited for walking upright. Proponents of ID would do well to acknowledge Darwin's revolution and accept natural selection as the process that accounts for the design of organisms, as well as for the dysfunctions, oddities, cruelties, and sadism that pervade the world of life. Attributing these to specific agency by the Creator amounts to blasphemy. (Ayala 2007, 159–160)

We can certainly agree that ID has a serious theodicy problem. Its proponents have not been entirely silent on the question (Dembski 2009), but I feel no desire to help them make their case.

The trouble is that as things stand Ayala has only pushed the problem back one level. Instead of wondering why God directly produces so many instances of poor design, we now wonder why He set in motion a process that inevitably leads to poor design.

Following its publication, Ayala's book was the subject of an exchange of papers in the academic journal *Theology and Science*. Several of the contributors (Behe 2008, Blackmore 2008, Moritz 2008; see also Doran 2009) raised precisely this issue. Ayala replied:

> Nevertheless, some would say the world was created by God, so God is ultimately responsible; God could have created a world without parasites or dysfunctionalities. Yes, others would answer, but a world of life with evolution is much more exciting; it is a creative world where new species arise, complex ecosystems come about, and humans have evolved. This account will not satisfy some people of faith, and many unbelievers will surely find it less than cogent: a Deus ex machina. But I am suggesting that it may provide the beginning of an explanation for many people of faith, as well as for theologians. (Ayala 2008, 190)

It seems rather callous to suggest that the excitement of evolution is sufficient mitigation for the appalling quantities of death and suffering it entails. The generations of scholars toiling before Darwin somehow managed not to be bored by the lack of creativity in nature. It would seem that a nonevolutionary world can still be plenty exciting.

I fail to see, then, how Ayala's suggestion advances the discussion at all. Identifying the suffering in nature as a side consequence of the creative process God employed only absolves Him of responsibility if we can show that a more benign process is not possible. But that is precisely the problem with which we began.

* * *

This brings to an end our brief tour of evolutionary theodicy. As with most theological questions, we can do no more than consider the arguments on offer and decide for ourselves what we find it plausible to believe. Considering the manifest weaknesses in the attempts we have considered, it is unsurprising that so many people find it impossible to think of Darwinian natural selection as the sort of creative mechanism a loving God would employ.

For myself, I can only say that I think Bertrand Russell was entirely correct in saying:

> It appears that during those ages which so troubled [geologist] Hugh Miller, when animals were torturing each other with ferocious horns and agonizing stings, Omnipotence was quietly waiting for the ultimate emergence of man, with his still more exquisite powers of torture and his far more widely diffused cruelty. Why the Creator should have preferred to reach His goal by a process, instead of going straight to it, these modern theologians do not tell us. Nor do they say much to allay our doubts as to the gloriousness of the consummation. It is difficult not to feel, as the boy did after being taught the alphabet, that it was not worth going through so much to get so little. This, however, is a matter of taste. (Russell 1997, 80)

25

From Catastrophe to Consummation

Returning to our tour of the museum, Adam's sin had gruesome effects. On separate placards, we learn about venom (and other natural toxins), death, disease, carnivores, nature "red in tooth and claw," scavengers, cosmic aging, poisons, conflict, weeds, burdensome work, and suffering. In each case the placard details the state of affairs both before and after the fall. Bible verses are produced, rarely from Genesis itself, to support the museum's view of how God intended things to be.

Representative is the placard for "Venom." On the left side, under the heading, "Before Adam's Sin," it reads, "No venom." This is supported by Isaiah 11:8 (KJV), "And the sucking child shall play on the hole of the asp, and the weaned child shall put his hand on the viper's den." The sign informs us that this is a "promise of a future time, similar to Eden."

This raises a question. God's creative work was completed during the events of creation week. If there was no venom at that time, but there is venom today, then where did it come from? The right side of the placard offers some suggestions:

> Though nothing harmed animals before Adam's sin, venom harms animals in the present. We do not know exactly how venoms first entered the world. Possibilities include
>
> - Changed use of chemicals (chemicals that once had non-harmful functions at the creation changed to venoms after the curse).
> - Revealed information (the potential to make venoms was built into the original creation, but not revealed until after the curse.)

The other placards display a similar attention to detail.

After a display detailing Cain's murder of his brother Abel, we come to an animatronic Methusaleh. "Something terrible is coming," he says. "No one believes what Noah says. But I can tell you, whatever God says is true."

Then we are standing inside a model of Noah's ark! Nearby, animatronic Noah warns folks about the coming judgment, while animatronic scoffers are stubbornly unimpressed. The placards explore the building materials and hydrodynamic properties of the ark, the logistical problems involved in its construction, and the meaning of the word "cubit."

The following room considers the details of the flood itself. Says one sign:

> Noah built the Ark, but God brought the animals. Every kind of land animal would have come, including dinosaurs and other animals that are now extinct. Since the flood was to be over a year long and animals leaving the Ark would have to repopulate the earth, it is likely God brought young adults. Being smaller, they would also be easier to care for.

One diorama shows the panicked unsaved desperately trying to escape the rising waters. The ark, meanwhile, floats safely by. The image is disturbing. The ease with which creationists summon forth threat and menace against those who demur from their perspective is one of their least endearing qualities. This theme is really brought home by yet another animatronic Noah. Before him is a panel of buttons, each labeled with a different question. If you press "Why did you build the Ark?" he answers, in part:

> He then gave me specific instructions on how to build the Ark and what to bring inside. It has been difficult, but God has provided for our every need. I am still amazed that I found favor in His eyes and that He made a way for us to escape this terrible judgment. How I wish that others, even my own brothers and sisters, would have changed their ways and obeyed God so that they too might have been saved. But now it is too late.

The equanimity with which Noah describes the slaughter of every human being on the planet is rather unnerving.

More than anything else, it is their literal understanding of Noah and the flood that makes young-Earth creationists seem foolish. Questions about natural history and the age of the Earth are highly abstract. It takes training and education to understand the magnitude of the scientific case against them. Moreover, I suspect that most of us have an instinctive sympathy for those willing to thumb their nose at authority, even when we believe the rebellion is misguided.

But the implausibility of the Noah story is striking upon even the most casual reading. Leave aside the sheer futility of God's destruction (humanity, after all, was just as sinful and corrupt after the flood as it was before). Never mind the question of where all the water came from and disappeared to. Pay no heed to the practical difficulties of constructing a vessel as large as the ark. Just consider this, from young-Earth geologist Andrew Snelling:

> If, as the preponderance of evidence shows, the "created kind" or *baramin* was possibly equivalent in most instances to the family (at least in the case of mammals and birds), then there would have only been about 2,000 animals on the Ark. On the other hand, the total number of species of mammals,

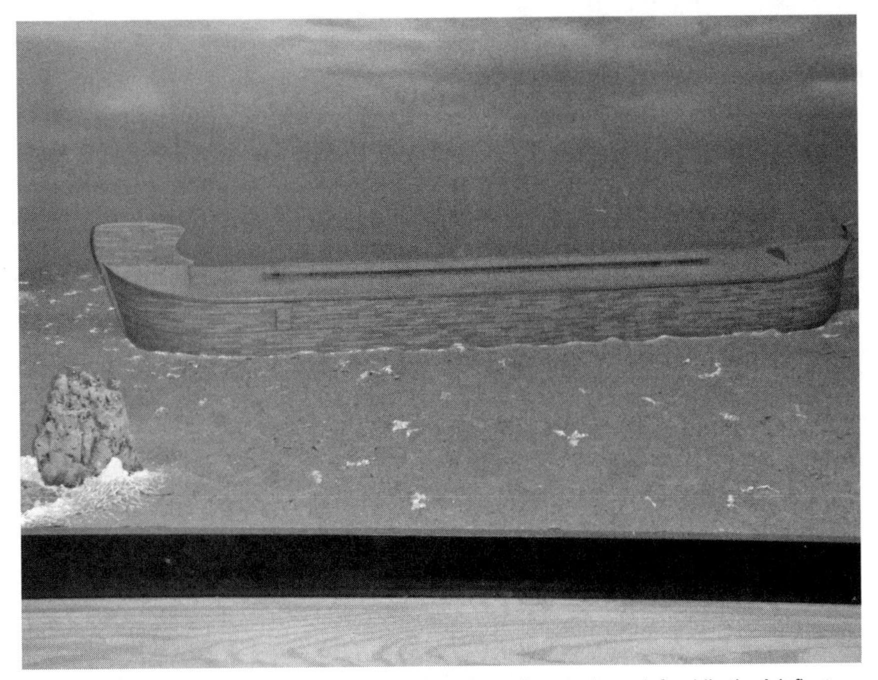

FIGURE 25.1 The unrepentant seek high ground on the rock at the lower left, while the Ark floats safely by.

> birds, reptiles and amphibians listed by [biologist Ernst] Mayr is only 21,000, so for all practical purposes one could say that 43,000 individual vertebrate animals would be the maximum number which would have needed to have been on the Ark. Therefore, it is most likely that the number of animals on the Ark was somewhere between these two extremes, and so [creation scientist John] Woodmorappe has suggested that if we adopt the genus as the taxonomic rank of the *baramin*, then approximately 16,000 animals would have been on the Ark, based on the land animals whose existence we know of, either as living animals or fossils. (Snelling 2009, 136–137)

As described in Genesis, the flood lasted more than a year. I ask you simply to consider floating in a small, enclosed boat with sixteen thousand animals for that length of time. The smell and the claustrophobia alone pose serious problems.

Creation scientists have not ignored these problems. They devote considerable effort to resolving them, and, as with everything else, they have their counterpoints to make. I can only say that, while I do make an effort to understand why people believe the things they do, in this case they truly seem to be defending the indefensible.

There followed numerous displays devoted to the basics of "flood geology," in which science is reworked to include the effects of Noah's flood. The displays here were considerably more technical than anything previously seen.

FIGURE 25.2 This exhibit of a dinosaur with a saddle symbolizes the creationists's belief that humans and dinosaurs coexisted.

After various exhibits devoted to the repopulation of the Earth after the flood, we come to the Tower of Babel. This event not only marks the confusion of human languages, but also the origin of all "human religions." One sign reads:

> At Babel, humans rejected God's plan, worshiping the creation rather than the creator, and following their own way rather than God's way. All human religions have followed their example, inventing myths to replace God's account of creation and Noah's Flood. They ignore God's warnings of judgment and His promises of blessing.

More alarming is yet another sign warning us of the evil fruits of human reason: "According to human reason, everyone does what is right in his own eyes. Once people abandon the authority of God's Word, there is no sure foundation for morality and justice in the world. Human reason can be used to justify evil of every sort." Accompanying this are pictures depicting some of those evils, such as racism, genocide, and abortion. With regard to racism, they quote Stephen Jay Gould, who wrote, "Biological arguments for racism may have been common before 1859, but they increased by orders of magnitude following the acceptance of evolutionary theory."[1]

Leaving aside the blindness involved in ignoring the role religion has played in fomenting evil, a nearby exhibit provides an ironic counterpoint. Under the heading, "Babel Explains Our Differences," appears the Bible verse, "Of the three sons of Noah the whole earth was overspread." Noah's sons are Japheth, Shem, and

Ham. A large diagram shows the descendants of Japheth settling in Europe, the descendants of Shem settling in Asia and the Middle East, and the descendants of Ham settling in Africa. This is highly significant in light of Genesis 9:20–27:

> Noah, a man of the soil, was the first to plant a vineyard. He drank some of the wine and became drunk, and he lay uncovered in his tent. And Ham, the father of Canaan, saw the nakedness of his father, and told his two brothers outside. Then Shem and Japheth took a garment, laid it on both their shoulders, and walked backwards and covered the nakedness of their father; their faces were turned away, and they did not see their father's nakedness. When Noah awoke from his wine and knew what his youngest son had done to him, he said, Cursed be Canaan; lowest of slaves shall he be to his brothers. He also said, Blessed by the Lord my God be Shem; and let Canaan be his slave. May God make space for Japheth, and let him live in the tents of Shem; and let Canaan be his slave.

The notion that the descendants of Ham were cursed, coupled with the notion that those descendants settled in Africa, was an entirely common way of justifying racism and slavery well into the nineteenth century (Haynes 2002).

It is an amusing juxtaposition to see a poster excoriating evolution for promoting racism next to a Bible-based notion that also promotes such thinking. As we shall see again in chapter 31, casual charges of racism against evolutionists are a regrettable commonplace of creationist argumentation. There is an irony in this, given that in other times and places creationists objected to evolution precisely because it promoted racial equality. In his classic work *The Mind of the South*, W. J. Cash writes, with reference to anti-evolution organizations in the South in the 1920s and 1930s, "One of the most stressed notions which went around was that evolution made a Negro as good as a white man—that is, threatened White Supremacy" (Cash 1941, 339). A proper understanding of evolutionary science shows that, on this one point, they were right (Gould 1984).

Moreover, in a study of beliefs among high school students in Arizona, it was found that more than 10 percent either "agreed" or "strongly agreed" with the statement, "The color of a person's skin depends on whether God favored or punished their ancestors" (Lawson and Worsnop, 1992). Creationism and racism have a long history of going hand in hand (McIver 1994, Moore 2001).

I do not raise these points to argue that creationism is either inherently racist or leads inevitably to racism. It isn't and doesn't. It's just that whenever I hear propagandists for creationism claim that evolution leads to racism, on the grounds that it's possible to find people in history who have used evolution to defend racist ideas, I am reminded of Jesus' admonition, "Why do you see the speck in your neighbor's eye, but do not notice the log in your own eye?" (Matthew 7:3).

Which brings us to the final stop on the tour, the film *The Last Adam*, providing a brief account of Jesus' life and death. When I had reached this point, a display on the wall informed me I had ten minutes to wait, so I found a vacant bench and sat down. There was certainly much to ponder in what I had seen. Some of the displays had been fun, but the incessant emphasis on wrath and judgment was a bit overwhelming.

I contemplated weighty questions. Like if maybe the creationists have a point about Genesis.

26

What Does Genesis Mean?

The young-Earth view certainly has the virtues of simplicity and clarity. Most people, upon noting that the days of Genesis are marked by the passage of an evening and a morning, and that they are sequentially numbered in a manner reminiscent of the days of the week, would conclude that twenty-four-hour days were intended. When they subsequently read, in Exodus 20:8–11, "For in six days the Lord made heaven and earth, the sea, and all that is in them, but rested the seventh day; therefore the Lord blessed the sabbath day and consecrated it," they would consider the case to be closed. In these verses the days of God's creation are specifically analogized to the days of our work week. If the author of Genesis 1 intended something other than twenty-four-hour days, then he went out of his way to be confusing.

Consider also the genealogies in Genesis. We are given a list of generations linking Adam to Noah (Genesis 5) and Noah to Abraham (Genesis 11). These lists include the precise ages at which various people bore their children. One wonders about the point of such detail if we are not meant to draw inferences regarding the time lapse between the creation and Abraham. These genealogies, let us be blunt, do not make for riveting reading. What possible purpose do they serve, asks a creationist, if we cannot take them at face value?

The fact is that if you hold the Bible at arm's length and only read the highlights you end up with an interpretation that is very much along their lines. You certainly do *not* get the impression that humans arrived billions of years after the moment of creation or that our appearance was the result of eons of evolution by natural selection.

I have suggested that the widespread and persistent acceptance of evolution in the scientific community is evidence, at least, that it cannot be dismissed as obviously absurd. A similar principle applies here. If you believe that the YEC approach is ludicrous, and not merely incorrect, then you really must explain how so many generations of Christians managed to share in such delusions.

Theologian Langdon Gilkey writes:

> To the Christian of the eighteenth and nineteenth centuries the biblical idea of creation probably meant something like this: "Not so very long ago (4004 B.C., in fact), God, having dwelt in splendid isolation for eternity, suddenly created in one series of momentous, instantaneous acts the whole present world. In this single miraculous series of events, centering somewhere in Mesopotamia, the Lord made in their present form all the kinds and species of things that were ever to be: the sun, the moon, and stars were given their places, our present seas, mountains, and valleys were formed by His direct power, the present species of plants and animals were made by His hand. Thus the whole world as we know it came to be, not by an age-long process of gradual development, but by the fiat of a fabulous artificer in six days of furious activity." (Gilkey 1965, 7–8)[1]

Gilkey was himself a theistic evolutionist (he testified for the evolution side in the 1981 *Arkansas* trial), and his contempt for the literal interpretation of Genesis is unmistakable. As a description of how most people used to understand the biblical text, however, his statement is surely correct.

Religion scholar Marcus Borg describes things similarly:

> Prior to the birth of modernity in the Enlightenment of the seventeenth and eighteenth centuries, however, the factual truth of Genesis was accepted in the Jewish and Christian worlds without controversy, even though its stories were not always read literally. There was little or no reason to question their factuality. Theology and science alike took it for granted that the universe was relatively young and that the earth and its continents, mountains, oceans and varieties of life were created in very much the same form in which we now find them. Common estimates of the time of creation ranged from 6000 BCE to 4000 BCE. (Borg 2001, 59)

Borg's assertion that there was "little or no reason to question their factuality" is very interesting. It suggests that the text itself does not provide such a reason. That is, we are not considering the psalms, which are obviously poetry, or Jesus' parables, which are obviously fictional stories meant to illustrate a point. Genesis 1 reads like an historical account. Old testament scholar William Brown writes, "Set beside the mystical poetry of the Egyptian traditions, Genesis 1:1–2:3 reads like a dispassionate treatise. By its own measure, Genesis resembles more an itemized list than a flowing narrative, more a report than a story" (Brown 2010, 33).

As recently as the seventeenth century, it was commonplace to use the historicity of Genesis as the basis for geochronology. James Ussher's determination that the world was created in 4004 B.C. is perhaps the most famous of these attempts, but there were many others. Geologist Patrick Wyse Jackson writes, with reference to the sixteenth and early seventeenth centuries:

> At about this time chronologies based on the sacred texts frequently started to appear in print although earlier chronologies were known, such as that published in *Cooper's Chronology* in 1560. Certainly the chronology published by Archbishop James Ussher (1580–1656) is the best known.... Consequently one might imagine that his was one of the few seventeenth-century estimates of the Earth's antiquity. This presumption is entirely erroneous: there were numerous chronologies besides Ussher's. (Wyse Jackson 2006, 16–17)

Wyse Jackson lists more than a dozen other estimates produced during this time, with dates of creation ranging from 3960 B.C. to 5233 B.C., and then continues:

> In 1809 William Hales documented 156 chronological estimates, and in 1861 Leonard Horner the geologist suggested that in fact the number was closer to double that estimate. These showed a range in time since the Creation of 6,500 years maximum to 3,600 years minimum duration. (ibid. 28)

It is interesting that no one working in this vein managed to deduce the Earth's great antiquity.

In presenting these excerpts I do not mean to imply a false unanimity regarding the proper interpretation of Genesis. Certainly many people in the history of Christian thought took portions of the text as figurative, especially with regard to the days of Genesis or the story of Noah. My point is simply that the modern creationist view was well within the bounds of mainstream thought, to the point of being dominant, for much of Christian history.

What you think the Bible means depends in part on what you think the Bible is. Therein lies the distinction between hermeneutics and exegesis. The former refers to the principles brought to understanding the text, while the latter involves your interpretation within that basic framework.

Young-Earth creationists have strong ideas about hermeneutics. They believe that God is the ultimate author of the Bible. Though He chose to work through human scribes, the words on the page are God's words. As YECs see things, throughout history God has communicated with humanity in a variety of ways—most dramatically through Jesus Christ, the Word incarnate. Recall John 1:1, "In the beginning was the Word, and the Word was with God, and the Word was God." God has also communicated directly to various human beings, as He did with Adam in the Garden of Eden. Within this framework the Bible is understood as "the Word" in written form (Wise 2002, 14–17). Seen in this way, the Word must be viewed as a coherent whole. It tells one consistent story from start to finish.

The Bible should also be seen as an expression of God's love.[2] Though horrified by our sin, He does not abandon us to the mercies of our corrupted natures. Among the purposes of the Bible is the transmission of factual information relevant to understanding our spiritual condition. These facts are anchors we can use against the onslaught of sin and temptation from the prevailing culture. Given this, does it not make sense that the Bible would begin with several chapters laying out the central realities of our existence? Does this really seem like the appropriate place for myths, or allegories, or poetry?

Having come this far, you can understand the importance of "perspicuity" in YECs' thinking. Why would God, in conveying the central truths of human existence, have expressed himself with anything but the utmost clarity? Ponder again the plight of the geochronologists. These were devout men who wanted nothing more than to understand God's Word. They devoted their life to such researches and were of unquestioned integrity and intelligence. But from a modern perspective it appears they were wrong about everything.

Human writers routinely manage to communicate their intentions through their words. Are we really suggesting God was unable to do likewise?

A common mantra among those seeking to discredit YEC is, "Genesis was never meant to be taken literally!" It is unclear what this means. The stories comprising the early chapters of Genesis are themselves the product of an evolutionary process, beginning as part of an oral tradition. It seems clear that attitudes toward these stories themselves changed over time. What started as a device for preserving the laws and traditions of a people endlessly on the move from various tormentors eventually became a repository of eternal and infallible wisdom. Bible scholar James Kugel writes:

> But by the second century B.C.E., all this was changing: for the Jews Scripture itself had become God's great book of instruction—no longer merely the record of events from the distant past of one people, nor prophetic oracles delivered to a specific audience, but words of eternal validity that were relevant, therefore, to anyone in any age. (Kugel 1997, 14)

Kugel goes on to describe four basic principles that were near universal to biblical interpreters of the time (ibid. 16–18). Paraphrasing, they were the following: (1) The Bible is fundamentally cryptic in that it might conceal great depths of significance beneath the surface meaning of the text. (2) The Bible is one great book of instruction, directly relevant to every aspect of daily life. (3) The Bible is perfect and perfectly harmonious. (4) All of scripture is divinely inspired.

The attitudes of the earliest people to encounter these stories is now lost to history, but it seems clear that, from a relatively early stage of their development, they were treated as the holy and inspired word of God. Kugel documents the opinions and concerns of the major second-century interpreters, and it is clear that these men did not regard the creation stories as purely allegorical.

Things become still more complex when we add the New Testament documents to our considerations. It has long been a tenet of Christian theology that the

Old Testament cannot properly be understood except in the light of the New Testament. For example, the story of Adam's fall takes on new resonance when pondered in the context of Jesus' life and death. The prophecies of Isaiah take on new significance when they are fulfilled by Jesus. Given this, it is not so clear that the attitudes of the ancient Israelites are normative for readers today. The Israelites were in the middle of a grand, unfolding story, one that would reach its climax with the appearance of Jesus. We might suppose they did not fully understand what they had.

Earlier I commented that you must hold the Bible at arm's length to find the YEC view compelling. The problem is that for all their emphasis on hewing to a plain reading, they routinely fail to apply their own principles.

Genesis 1:6 (KJV) reads, "And God said, Let there be a firmament in the midst of the waters, and let it divide the waters from the waters." The Hebrew word translated as "firmament" refers to a thin metal sheet that has been pressed out or banged flat. The New Revised Standard Version and New American Version translate the Hebrew word as "dome," while the New International Version uses "expanse." This understanding is consistent with the thrust of the verse, which describes the firmament as a physical divider between the waters below and the waters above. Rain would then be interpreted as the firmament temporarily opening, thereby allowing some of the waters above to fall to the Earth below.

This is the first of many verses suggesting a flat Earth covered by a dome-shaped firmament, with God in His heaven above and the underworld below. We know, from other extant documents, that this was the commonly accepted view of the time. No such physical divider exists, of course, which is why YECs interpret the "firmament" to refer to space in general. This, however, is not a natural or plausible reading.

The problems continue in Genesis 2. It is commonly said that the first two chapters of Genesis contain separate, and contradictory, accounts of creation. For example, whereas the first chapter depicts animals being created before humans, the second chapter reverses that ordering. Creationists bristle at such charges, arguing instead that the two accounts focus on different aspects of the same story. The second chapter, they argue, simply does not depict a chronological sequence of events, thereby dissolving the contradiction.

This is a very dubious claim, but let us leave that aside. The undeniable fact is that the writing style changes dramatically between the two stories. It strains credulity to think that a single author striving for narrative continuity would have expressed himself in such a way. If the intent of Genesis 2 were to elaborate on certain aspects of Genesis 1, there would be some clear indication of that in the text. Instead we seem to have different stories written by different authors.

This impression grows in Genesis 3:17, where we encounter Cain's wife for the first time. Every reader, upon encountering this verse, has found himself flipping backward to find what he or she had missed. A single author would surely have realized that he had just mentioned a character that had not previously been introduced. Creationists are touchy about this (there is, in fact, an exhibit in the Creation Museum addressing it specifically). They point to Genesis 5:4, in which we learn that Adam and Eve had children, specifically daughters, who are not explicitly named in the text. Presumably Cain married one of those daughters. They will then explain to you, at length, why we should not be bothered by this particular example of incest.

I invite you to take up that topic with them, but for me it is beside the point. The author of Genesis 3 plainly expected us to be familiar with Cain's wife, but at no point is her existence explained in the story. The clear implication is that the story was not originally written with its current location in mind. We seem to have a story removed from its original context and placed, awkwardly, somewhere else. That impression deepens after the murder of Abel, when Cain is cursed by God. Genesis 4:14 reads, "Today you have driven me away from the soil, and I shall be hidden from your face; I shall be a fugitive and a wanderer on the earth, and anyone who meets me may kill me." If the entire population of the Earth consisted of Adam, Eve, and a few of their offspring, one wonders of whom Cain was afraid. Several other verses confirm the view that the story's author presupposed a well-populated Earth (Moberly 2009). This tends to confirm the impression that the story of Cain, as we now find it, was not originally part of a continuous narrative with the creation accounts.

The creationist response is based on the long life spans described in Genesis. The events in Genesis 4 may have taken place several centuries after Cain and Abel were born, thus providing ample time for the "other sons and daughters" to produce a substantial population. If this is correct, though, we have another instance of poor writing. The fourth chapter opens with the birth of Cain and Abel, and then proceeds directly to the details of Cain's fratricide. There is no mention of a long duration during which a civilization sprang up. You would expect the author to mention such a thing, especially given its relevance later in the story.

These are just a few items. There are many others. All pose severe, I would say insurmountable, challenges to the young-Earth interpretation. Even a fairly casual reading of Genesis reveals it to be a patchwork of stories from different cultural traditions, imperfectly edited together.[3]

Such objections notwithstanding, a thought experiment suggests itself. Suppose that science had discovered that the Earth was roughly ten thousand years old. Suppose further that the fossil record confirmed the instantaneous creation of all animals, and that biologists found a variety of genetic codes in nature, distributed in a manner consistent with a reasonable notion of "created kind." Would anyone consider that an embarrassment for the Bible? Would anyone today be arguing that Genesis *obviously* was not meant to be taken literally or to instruct us in science? Or do you think instead the objections raised in the last section would then be seen as nitpicks?

* * *

There is no shortage of people, from biblical scholars to more moderate evangelicals, who demur from the young-Earth view. We might wonder, then, where they locate the flaw in YECs' reasoning. Is the problem with their exegesis or with their hermeneutics?

Finding fault with their exegesis are a variety of approaches that, no less than young-Earth creationism, insist on the inerrancy of the text. One such is the day-age view, in which the "days" of Genesis 1 are understood to be long periods of time (Snoke 2006). In this way the Bible is construed as taking no clear stand on the age of the earth, thereby avoiding conflicts with geology. The genealogies in Genesis 5 and 11 are seen as incomplete. The events of creation week are still understood as historical, however, and the leading modern advocates of this view are very clever about aligning these events with what science has to say.

While the possibility that the Genesis days are not twenty-four hours long has a history going back to St. Augustine, and has always had defenders among Christians, the day-age view specifically was a response to geological advances in the early nineteenth century. Long before Darwin or radiometric dating, the vast layers of accumulated sediment revealed by mass digs had rendered the traditional biblical time frame untenable. Reinterpreting that time frame was seen as a device for preserving the central truths of Genesis while discarding secondary details (Rogerson 2009).

Regrettably, the day-age view is exegetically implausible. Theologian David Wilkinson writes:

> Although *yom* is used as a longer period of time elsewhere in the Bible, it is highly unlikely that the writer of Genesis I was using it in that way. This is shown by the days being used as part of a week and with the emphasis on "evening and morning." Further, the agreement between the general order of the biblical creative acts and the fossil record is not exact, so that trees appear before marine creatures and evening and morning appear before the sun and moon. (Wilkinson 2009, 136)

Quite right. It really seems far-fetched that a writer wishing to communicate long periods of time would write anything remotely like what is found in Genesis 1. Moreover, chronological conflicts are the least of the problems, since, as Wilkinson notes, the biblical sequence of events is not at all consonant with the findings of science.

A second possibility is known as the framework hypothesis. Its central insight involves a curious relationship between the first three and last three days of God's creative period. Philosopher Roy Clouser explains:

> If we keep the text's religious focus in mind while examining its organization, it then looks even less like an encyclopedic source of scientific information. It speaks of God's creating as taking place in six days, and reports the days as follows: Day 1, God separates light from darkness; Day 2, God separates sea

> from atmosphere; Day 3, God separates land from sea and creates plants; Day 4, God creates the sun, moon and stars; Day 5, God creates sea life and birds; Day 6, God creates plants, animals and humans. There is an obvious correspondence here between days 1, 2, 3 and days 4, 5, 6. Day 1 speaks of the difference between light and darkness as the planned precondition for the appearance of the sun, moon, and stars on Day 4. Day 2 offers the separation of atmosphere from sea as the precondition for the appearance of sea life and birds on Day 5. And the appearance of dry land and plants on Day 3 is the precondition for the creation of animals and humans on Day 6. This match-up of the first three days with the second three is too striking and essential a feature of the account to be accidental. But if it's not accidental, it shows two very important things. The first is that the intent of the text is to supply a *teleological*, not chronological account of God's ordering of the earth and its life forms. The order here is an order of purpose, not time. (Clouser 2001, 517)

This is terribly clever, but it seems even more implausible than the day-age view. Does not the magnitude of the conclusion strike you as out of all proportion to the evidence presented? I suppose it's possible that the parallel triads are proof that we are not dealing with a chronological sequence. But it is also possible that it suited God's purposes to create in this order. It is, after all, a perfectly sensible order in which to create things.

Clouser continues:

> The second thing the correspondence supports is that the "days" are to be taken as a literary framework for speaking of God's creative activity. The reason is both obvious and simple when the connection of this account to the Sinaitic covenant is kept in mind: since that covenant requires that humans work six days and rest on the Sabbath, God's work is represented as done in the same way. God's work thus becomes a model for how humans are to work. (ibid. 517–518)

Again, the conclusion has little relation to the evidence. I fail to see the connection between the two triads and the conclusion that the six days are a literary framework imposed on God's creative activity for the purpose of simulating a work week. In Clouser's telling, God wished to provide humans with a model for their own work. God might simply have organized His activities accordingly, precisely as the creationists claim and a plain reading of scripture supports. Instead, it is argued, He organized His work in a manner not specified in scripture, and then He imposed upon whatever He did the structure of a familiar work week. The primary clue that He behaved so remarkably is the parallelism between the two triads. This is not reasonable.

Still another approach will be familiar to anyone who has found himself explaining a difficult notion to a young child. Inevitably we simplify things, perhaps to the point of not being strictly accurate, so that the child may understand.

We could imagine that God faces a similar problem in dealing with us. Biologist Denis Lamoureux explains:

> Instead, when the Holy Spirit inspired the writers of the Bible, He *accommodated*. He lowered Himself and met them and their readers at their level. In the same way that Jesus meets us wherever we happen to be, so too the Holy Spirit inspired the authors of Scripture by using their terms and concepts about the natural world in order to reveal as effectively as possible Messages of Faith. (Lamoureux 2009, 65; see also Collins 2003)

"Inerrancy," on this view, refers only to the Bible's spiritual teachings. The text contains bona fide errors of fact, but only because God found it necessary to convey His teachings through the idioms and understandings of the time.

The principle here is certainly sound. It is hardly uncommon to use fictional stories as devices for communicating greater truths. Jesus Himself routinely taught through parables, a point often raised in support of this general view. To the extent that this approach allows us to deal honestly with the text, and does not try to pretend either that the biblical scribes anticipated modern science or that they wrote a complex cipher in need of elaborate decoding, it represents an improvement over what we have seen thus far.

But that is where the good news ends. To an even greater degree than with the day-age view and the framework hypothesis, we can be confident that this interpretation is incorrect. Why is there not a single verse in Genesis warning us that the text is an accommodation to limited human understanding? Where is the biblical equivalent of "Let me put this in terms you will understand"? Just consider the harm God's accommodations have caused. Think of all the Christians driven to anti-science views and all the skeptics turned off by the Bible's manifest errors. God must surely have known this would happen. Is it credible to think God would communicate the central truths of human existence in a manner confusing to nearly everyone who read it for centuries? Consider also that the account in Genesis 1 is not an oversimplified version of what really happened. From the perspective of modern science, it appears to be a complete fabrication. Accommodating oneself to an audience with limited understanding is one thing, but simply inventing a story from whole cloth is quite another.

Nor should we pretend that the Bible's scientific assertions can be cleanly separated from its theological teachings. One example is that Adam is treated as an historical figure throughout the New Testament, and his transgressions are explicitly presented as the reason Jesus' sacrifice is necessary. If we are to dismiss Adam as mythical, as modern science forces us to do, then what does that imply about Jesus? And if even these most central truths of the Christian faith must be reimagined in the light of modern science, then upon what can we ground our faith?

Perhaps we should reconsider our hermeneutics. If the premise of verbal inerrancy leaves us without a plausible exegesis, then maybe the premise is incorrect. No doubt some will consider this heretical, but it is not so clear that they are right.

Is there not something bizarre in the notion of God coauthoring a book? Does that really cohere with what we know of God's interactions with humanity? God's love is persuasive; he leads us through longings in our heart and not through gaudy, unambiguous displays. We see this most powerfully in the example of Jesus. God might have come down on a flaming chariot and revealed Himself to all. Instead, He accepted the limitations of humanity and provided us an example to follow. It was an example so powerful that people were led to abandon their former lives after brief contact with Him. At no point does Jesus say anything like, "I am God, cower before Me." Instead He is coy, asking questions like, "Who do you say that I am?"

So I ask again: Is it sensible to think that God would act by first inspiring a human scribe and then so constraining his mind and his hand that only God's words appear on the page? How is that consistent with this God of persuasion and self-limitation? Far more plausible is that people have experiences of God, moments of insight into His reality and boundless love, and are so moved they must write them down as best they can. In our discussion of religious experiences, we found that people who believe they have had them agree on the insufficiency of words to express what occurred. So it is with the biblical authors. They have provided us with powerful accounts of their experiences, imperfectly expressed in the idioms of the day.

Langdon Gilkey explains:

> It has certainly been an added complication that the Scriptures which contain this early story of creation were regarded as verbally inspired. Thus every literal statement of the story, whether the content of the statement concerned "scientific" subjects or religious affirmations, was regarded as unequivocally true.... God's revelation to men has come in and through the historical events to which the Bible witnesses; it has not come through all the words, stories, and statements of which the Bible as a book is composed....
>
> The words of the Bible, therefore, point us to the revelation which begins in the life of the Hebrew people and culminates in the event of Christ; they are not themselves the revelation. Consequently, the stories of which the Old Testament is filled are not "revealed stories"; what is revealed is the divine activity within history to which these stories witness. (Gilkey 1965, 27–28)

Marcus Borg is even more succinct: "They are not God's stories of the world's beginnings; rather, they are ancient Israel's stories of the world's beginnings" (Borg 2001, 62).

From the perspective of Christianity, I would think this model is entirely plausible. More plausible, in fact, than familiar notions of verbal, plenary inspiration. Of course, following Gilkey requires discarding certain traditional teachings, but perhaps that is good riddance to bad rubbish.

A believer might object that being so cavalier about Genesis leaves us without a reliable guide to which parts of scripture can be treated as historical and which are merely allegorical. If Genesis is not historical, then perhaps the parts about Jesus aren't either. My reply is that you should recall whatever it was that convinced you of the Bible's veracity in the first place. Was it *really* the scientific accuracy of Genesis? Was *that* what led you to devote your life to Christ? If it was, then I agree that we have a problem. If instead your faith was based on experience or on the powerful testimonies in the New Testament, then I fail to see what has been lost by adopting Gilkey's approach.

Christians have options when it comes to interpreting Genesis. That is why, for all the genuine difficulties evolution poses to Christianity, I do not regard conflicts with scripture as one of them.

I would hasten to add, however, that there is another possibility. We could view the stories in Genesis as entirely human productions. If they seem scientifically naive to us today, that is because they were written by scientifically naive people.

Bible scholar Daniel Harlow has compiled an impressive list of commonalities between Genesis and the creation myths of other civilizations from that time (Harlow 2010). Almost every major theme, from an initial garden paradise to clothing as a symbol of civilized life, can be found in stories from other Mesopotamian civilizations. Harlow himself uses this observation as the basis for an argument that Adam and Eve were meant as purely literary, not historical, figures, but we are hardly forced to that conclusion.

Our brief investigation has revealed copious evidence of editing and internal stress in the Genesis narratives. We find their major themes were common in stories of the time, adapted to fulfill the specific religious and political needs of the ancient Jews. The Bible's forays into science are frequently fallacious. What part of this suggests the text was divinely inspired? Which part of the text is mysterious under the assumption that it was written by noninspired human beings?

I spoke earlier of my bafflement over religion. This is another manifestation of it. To me it looks like people are working awfully hard to find meaning in Genesis that just isn't there.

27

Theological Phlogiston

After the film *The Last Adam,* I bought an ice cream sandwich from a snack bar outside the theater. It was tasty and refreshing and made me feel better. I brought it outside and walked around the garden. Eventually I found a shaded pavilion and sat down for a big think. So ended my tour of the museum.

What would you have me do at this point? Should I rant and yell? Find some especially clever put-down for the whole thing? Maybe I should wax eloquent about all of the better uses to which twenty-seven million dollars could have been put. Uses, mind you, that do not involve brainwashing children into a worldview that was already implausible two centuries ago. Shall I lament the failures of an educational system that has led so many to be ignorant of basic science and history? Hold forth on the infuriating persistence of American anti-intellectualism? Lecture about the death of the Enlightenment?

I *could* do all those things. At various times I *have* done all those things. But at that particular moment I was not in the mood. My feelings were not so much anger, but more like defeated helplessness. It's like the feeling you get when you see an ant on your kitchen counter and think, "That's not so terrible," until you open your cabinets and see dozens more. Or the moment when, after hearing a strange noise from your car, you go from thinking something needs to be tightened to understanding you need a new transmission.

I try to imagine the planning meetings that went into this little operation. The teams of organizers who must have spent years arranging it all. The ease with which they raised the money and the flocks of visitors who have made it a success. Substantial imagination went into designing the exhibits and films, and I am brought close to tears by the thought of such talent being wasted in this way.

Then I ruminate on those moments when I allow my mind to wander, and I lose myself in the museum's exhibits, in much the same way as one loses oneself in the pages of a good novel. The moments where I agree not to look too closely at the details and to forget what I know about these topics. There are times when I achieve the requisite suspension of disbelief and, if only briefly, manage to enjoy myself. But then I think of how it looks to all the children who, if their parents are successful, will know *only* this, and suddenly I am no longer enjoying myself.

Isaac Asimov once wrote, "To those who are trained in science, creationism seems a bad dream, a sudden coming back to life of a nightmare, a renewed march of an Army of the Night risen to challenge free thought and enlightenment" (Asimov 1983, 6). He's right. That's exactly what it is.

Looks like I did have a rant in me. It's just that I also have moments when I understand why creationists are so frustrated. Consider this statement, from Marcus Borg:

> Given the richness of meaning that a historical-metaphorical reading of Genesis reveals, the creation stories strike me as profoundly true. Critical thinking leads to an understanding of why the details of Genesis are as they are and also makes clear that their truth is not to be understood in literal, factual terms. Rather, their truth is expressed in the nonconceptual language of myth and metaphor, and no particular reading can exhaust their meanings.
>
> But I can hear the truth of their central claims. "This"—the universe and we—is not self-caused, but grounded in the sacred. "This" is utterly remarkable and wondrous, a Mystery beyond words that evokes wonder, awe, and praise. We begin our lives "in paradise," but we all experience expulsion into a world of exile, anxiety, self-preoccupation, bondage, and conflict. And yes, also a world of goodness and beauty: it is the creation of God. But it is a world in which something is awry. (Borg 2001, 80–81)

I do not wish to be churlish, but as profound truths go these seem pretty banal. We hardly need Christianity to direct us towards the awesomeness of existence, or to the distance between what the world is and what it could be. Such observations are commonplaces of daily life. It used to be, however, that Christianity made genuine contributions to our understanding of why things are as they are. In the story of Adam and Eve, and the closely related doctrine of original sin, we had cogent explanations for why humanity so often fell short of its highest ideals. But those explanations are no longer tenable. Today, such knowledge as we have regarding human nature come entirely from sources other than religion.

This is part of the reason creationists are so unimpressed with those who would try to reconcile Christianity with evolution. Too often the Christianity with which they are left just seems toothless. Here is another example, this time from physicist Karl Giberson:

> We must concede that the acid of evolution has indeed eaten away the literal part of this story [of Adam and Eve], but I would argue that the most important part of the story remains untouched.

> The idea at the center of the fall is human sinfulness. Human beings are sinful creatures, and many of us are really quite dreadful. Even the best of us dare not lay claim to anything approaching perfection. . . . The classic story of the fall is best understood as a powerful statement that we are, when all is said and done, sinful creatures. (Giberson 2008, 12)

I strongly disagree that the most important part of the story remains untouched. Reducing the story of the fall to a bland observation about human sinfulness trivializes it to the point of irrelevance. Once again, we hardly needed Christianity to tell us that humans often behave poorly, and evolutionary biology has far more light to shed on why we are as we are than the Bible does.

Moreover, the story of the fall is presented in the New Testament explicitly as a factual explanation for humanity's sinful state. Recall Romans 5:18, "Therefore just as one man's trespass led to condemnation for all, so one man's act of righteousness leads to justification and life for all." Given this, it is unsurprising that many believe something important has been lost in discarding the historicity of Adam and Eve.

There is a substantial literature devoted to reimagining the doctrine of original sin in the light of evolution. Here is one possibility: Our "sinful nature" can today be understood as our inheritance from the selfishness inherent in the evolutionary process. While that inheritance certainly does not come to us through a single man, it does, in fact, come from the universal common ancestor. Paleontologist Daryl Domning defends this view, writing:

> What I have sought to show is that the *overt selfish acts* which, in humans, demonstrate the reality of original sin (by manifesting it in the form of actual sin) do indeed owe their universality among humans to natural descent from a common ancestor. This ancestor, however, far from being identifiable with the biblical Adam, must be placed in the very remote past, indeed at the very origin of life itself. It was the common ancestor not only of humans but of all other living things on Earth as well. However, it is not this ancestor itself that is of real interest, but the "natural descent" that proceeded from it: the very nature of physical life and the process of natural generation, which are governed by natural selection and the selfish behavior it requires. (Domning and Hellwig 2006, 140–141)

There is no "fact of the matter" regarding the proper understanding of original sin. If Domning, or anyone else, finds it fruitful to view things this way, then it is not for me to tell them they are wrong. Under Domning's view, however, one can reasonably wonder whether the concept of original sin is making any contribution at all to our understanding of the human condition. It seems that we are just taking a body of knowledge provided by science and attaching a Christian label to it.

Theologian John Haught offers a different take on how to understand original sin in the light of modern science:

> Other frameworks would understand it differently, but in this setting original sin means that each of us is born into a still unfinished, imperfect universe where there already exist strong pressures—many of them inherited culturally over countless generations—for us to acquiesce in an indifference to God's creative cosmic aim of maximizing beauty. Original sin consists of all the forces that lead us away from participation in this most essential and vitalizing pursuit. (Haught 2000, 138)

Elsewhere he specifically cautions against the sort of view supported by Domning:

> Moreover, even though the potential to do evil is already a part of our genetic makeup, it is theologically inappropriate to identify original sin simply with the instincts of aggression or selfishness that we may have inherited from our nonhuman evolutionary ancestry. Even though these tendencies are part of our evolutionary legacy, the substance of "original sin" is the culturally and environmentally inherited deposit of humanity's violence and injustice that burdens and threatens to corrupt each of us born into this world. (Haught 2000, 139)

We have certainly come a long way from the traditional understanding of original sin, so much so that the name itself now seems inaccurate. Gone is any notion of a specific sin committed by Earth's original people. In reply I can only wonder again what original sin is contributing to our discussion. An atheist, no less than a Christian, is fully capable of bemoaning the societal forces that prevent humanity from living up to its potential. If references to original sin suggest nothing more than an acknowledgment of these forces, then I fail to see what is gained by preserving the religious terminology.

Perhaps we should not be hasty in discarding a literal Adam and Eve. Biologist Denis Alexander suggests an imaginative possibility. (To clarify what follows, he presents this as the third model out of five possibilities, and later makes clear it is the model he prefers):

> According to model C, God in his grace chose a couple of Neolithic farmers in the Near East, or maybe a community of farmers, to whom he chose to reveal himself—so that they might know him as a personal God. Model C also draws attention to the representative nature of 'the Adam', 'the man', as suggested by the use of the definite article in the Genesis text as discussed above. 'The man' is therefore viewed as the federal head of the whole of humanity alive at that time. This was the moment at which God decided to start his new spiritual family on earth, consisting of all those who put their trust in God by faith, expressed in obedience to his will. Adam and Eve, in this view, were real people, living in a particular historical era and geographical location, chosen by God to be representatives of his new humanity on earth, not by virtue of anything they had done, but simply by God's grace. (Alexander 2008, 236–237)

I regret that I lack the space here to discuss the minute, but unpersuasive, scriptural arguments Alexander uses to support this idea. I think, though, that we are entitled to be suspicious of an interpretation that suddenly discovers in the text of Genesis 2 a previously unsuspected population of Neolithic farmers. A story in which Adam is formed from the dust of the ground, with Eve then created from one of his ribs, is quite different from one in which Adam and Eve are chosen by God from among a large group of people. If Alexander is correct, then we must find some reason for believing that what certainly appears to be the former is actually symbolic of the latter.[1]

There are, I would think, theological problems as well. The whole model of "federal headship" is deeply problematic in this context. For what reason would God reveal Himself only to two people? If the point was to urge humanity, recently evolved from ape-like ancestors, to a higher state of spiritual experience, then why not reveal Himself to everyone?[2] Moreover, the problem of discerning the mechanism through which Adam's sin was transmitted to all of his descendants pales in comparison to that of finding the justice in punishing everyone for the sins of one couple. After the expulsion from Eden, were Adam and Eve's fellow Neolithic farmers made aware of the reasons for the world's novel state of fallen decrepitude? One suspects the first couple was not invited to many parties.

Alexander writes:

> Let us reiterate: of course the point with this model is not that the model itself is found within the Genesis text—it is not. The idea is to generate a working model that will explore the possible 'narrative behind the narrative', the events in human history that might at least be 'consistent with' the Genesis theological account. (Alexander 2008, 241)

Since Alexander has granted that his model is not found within the Genesis text, we can reasonably ask whether he is simply reading into the text that which is not there. Creationists, of course, would object to the whole approach of trying to make our understanding of Genesis cohere with our understanding of human history. They would wonder why we should be looking for a narrative behind the narrative, when a plain reading of the story seems to make perfect sense.

In this case, I think they have a point.

Those of us who enjoy reading typically have certain works of literature to which we consistently return. For me it is Victor Hugo's novel *Les Miserables*. If you can forgive the cliché, I find something new every time I read it.

Especially powerful is the character of Inspector Javert. The novel's main plot involves his relentless pursuit of Jean Valjean, a convict who broke his parole after a religious conversion compelled him to reform his life. Javert's obsessive devotion to the law, which he sees in religious terms, blinds him to everything except Valjean's ultimately trivial infraction. He is so unbending and inflexible that, when circumstances force him to realize the error of his ways, he is unable to change. After ten pages or so of heart-wrenching inner monologue, he commits suicide.

Every generation and every locality has its Inspector Javerts. In twenty-first-century America I see him manifested in many of the creationists I have met. Were I living in America in the 1950s, I would have seen him in the mania of Senator Joseph McCarthy. In early twentieth-century Russia it was Trofim Lysenko, whose delusional ideas set Russian science back decades and destroyed lives. In each generation the story is reinterpreted anew; that it persists in remaining relevant is evidence of the novel's greatness.

The Bible could be read with a similar attitude. It is a vast book with enough depth to keep us busy for several lifetimes. Even an ardent secularist like myself can find much to admire. The stories of Joseph and Jonah have much to teach us, even if we do not believe in God. They are also good yarns. The ruminations on suffering and futility found in Job and Ecclesiastes are as powerful and relevant today as they were when they were written. By reading them we feel connected to people alive thousands of years ago. The minutiae of our daily lives might change from one generation to the next, but our existential quandaries remain constant.

An inventive way of applying this attitude to Genesis 1 is provided by William Brown. He writes, "How do persons of faith living in a scientifically informed world appropriate the ancient creation traditions of Scripture? Appropriation involves not just a new way of interpreting the text but a new way of living it" (Brown 2010, 15). Later he elaborates:

> I view the text through *my* lens. To the text I bring my concerns and convictions, my filters and prejudices. I come to the text not as a blank slate but as a reader informed by science and shaped by my culture, which in turn shapes my interpretation, even my translation. The text is not a container of meaning waiting to be unlocked and opened, but an object of focus with which I interact and whose meaning emerges only by my interaction with it. (ibid., 16)

Toward that end he finds "virtual parallels" between modern science and biblical faith. The Bible's teaching of "creation by differentiation: from formless 'chaos' to order," is parallel to the modern scientific understanding of the "entropic rise of complexity: from uniformity to structure." In the biblical teaching of "life 'filling' various domains," we see a parallel to the scientific understanding of a "biological drive to 'fit' various environments." Brown is quite ingenious at this, and he produces an extensive list of such parallels from which these two examples were drawn (ibid., 241). Note, however, that he is not claiming the biblical authors anticipated modern science in any way. That is why he calls them "virtual" parallels and emphasizes the role of his prior beliefs and convictions in interpreting the text.

Those who find meaning in this sort of analysis are welcome to it, but I really must wonder about the point of the exercise. Permitted Brown's level of poetic license and creative interpretation, I could easily make the *Enuma Elish* seem insightful.[3] I would further note that more conservative theists are likely to find

Brown's approach not just unsatisfying, but actually disrespectful. For them the text is, indeed, a container of meaning. It is not a volume whose words are to be manipulated casually.

You might wonder why I have belabored these points. What's it to me how other people interpret the Bible? If others find it helpful to think in terms of original sin, or to read the Bible in creative new ways, then that's an expression of their personal faith. Who am I to criticize?

To the extent that we really are talking about matters of personal faith, I am happy to stay on my side of the line. The trouble is that all too often the people advancing these views are unwilling to stay on theirs. Endlessly we hear calls for dialogue between science and religion. Denis Alexander titles one of his chapters, "Who Were Adam and Eve? Genesis and Science in Conversation." He writes: "We have now briefly surveyed the Bible's teaching about Adam and Eve, and given a highly compressed account of our current understanding of human evolution. We are therefore ready to address the question, 'What relationship, if any, might there be between these two accounts?'" (Alexander 2008, 234). William Brown writes:

> If theology is, to quote St. Anselm, "faith seeking understanding" and science is a form of understanding seeking further understanding, then theology has nothing to fear and, in fact, much to learn from science. Theology cannot advance the scientific quest for the underlying constituents of matter and the physical nature of causation. Science, in turn, cannot lay claim to know God and God's purposes. Both disciplines represent independent fields of inquiry. But, I ask, does their independence preclude cross-disciplinary conversation? Because both seek truth each can learn from the other, especially theology from science. (Brown 2010, 8)

It would seem, then, that something more than just personal beliefs are at stake. The implication of describing Genesis and science as engaging in conversation is that each has something to say to the other. Likewise for saying that science and theology are both truth-seeking enterprises, each of which can learn from the other. Scientists, as scientists, are apparently expected to take an interest in what the theologians are up to. This presents a problem, because for the life of me I cannot see how theology intends to hold up its end of the conversation.[4] I will grant that science and theology both seek truth, but it is unclear, to put it gently, that theology has investigatory methods that are up to the task. I will likewise grant that science cannot lay claim to know God and His purposes, but what reason can be given for thinking theology can?

In the seventeenth and eighteenth century it was commonly thought that there existed a firelike element known as "phlogiston." Highly flammable materials were those that contained a large amount of this substance. Upon burning they became "dephlogisticated." Left behind was the "calx," which was thought to be the true essence of the material. The phlogiston theory was eventually refuted by careful experimentation. Today combustion is understood as a chemical reaction between oxygen and some sort of fuel. Highly flammable materials are those best able to engage in this reaction.

Scientists no longer talk much about phlogiston, unless they happen to be teaching a course in the history of science. It seemed like a good idea for a while, but has now been discarded in favor of more promising approaches. How do you suppose scientists would react to someone who argued as follows:

> Many scientists, apparently believing that newer is always better, have seen fit to scorn the accomplishments of their forebears with regard to combustion. It is certainly true that the old understanding of phlogiston as an actual, physical substance is today happily abandoned. It was the product of an age all too eager to think that only the material was real. A more modern understanding holds that phlogiston refers to the exquisite dance between fuel and oxygen, as described so brilliantly by modern chemists. Phlogiston is a process, not a substance. We are certainly grateful to those who contributed to this new understanding of the concept, and look forward to further engagement between modern science and older wisdom.

I suspect most scientists, after rolling their eyes, would tell him he is welcome to define his words however he likes, but that there is nothing to be learned from his reinterpretation. But if the phlogiston-defender then started accusing scientists of arrogance and dogmatism for being unreceptive to his ideas, the relationship might quickly grow chilly.

That is how so much theological writing seems to me. If you want to redefine original sin, or summon forth strained interpretations of Genesis to reconcile evolution with Adam and Eve, then go right ahead. But please do not pretend that this represents some convergence of ancient wisdom with modern understandings. This is not science and religion in conversation. This is science telling it like it is, and religion trying desperately to catch up. After science has dutifully applied its methods, over the course of centuries and frequently in the face of religious objections, you do not get to redefine your words and pretend that religion had the answers all along.

As a nonbeliever there are certain words that do not come naturally to me. Words like *holy, worship, faith, sacred, prayer, numinous, divine,* and perhaps most of all, *transcendent.* When I hear people use such language, it usually just sounds pretentious and overwrought to me. I do, however, get occasional glimpses into what "transcendent" might mean. What other word adequately captures the gloriousness of humanity's journey from frightened and primitive beginnings to ever greater understandings of the world?

Would you really like to know how to honor scripture? You do not do it by burdening the ancients with notions of infallibility, or by acting as though their simplest thoughts were expressed in poetry and metaphor, or by twisting their plain words into a form more consonant with modern science. You do not honor them by pretending they were possessed of special insight. They were just people, no different from anyone today, doing their best to make sense of their world. Reading their literature instantly connects you to fellow human beings far removed in space and time, not because of the answers they provided, but because of the questions they asked.

You honor scripture by seeing it as one link in a long chain. The creation account in Genesis 1 is, in a very practical way, a great advance over the other origins stories of the time. A world superintended by gods whose quarrels spilled into the earthly domain, gods we might hope to propitiate but could never hope to understand, is far different from a comprehensible world created for humans by a God of love. The future development of science owed much to this advance.[5] It is in that simple fact, and not in antiquated notions like original sin or Adam and Eve, that we find what is good in Genesis. How dishonoring it is to the ancients to focus on the errors in their details over all that was correct in their big picture, and over all that was admirable in the depth of their thought.

For a young child, crawling is an important advance, but once he learns to walk we do not long for him to return to the ground. By all means lambaste the creationists for their dogmatism and idiosyncratic exegesis. Just take the precaution of being sure that your approach represents an improvement.

28

Why I Love Being Jewish

That, however, is hardly the end of the story.

To judge from the museum's exhibits, you would think religion was primarily about the confident assertion of factual claims—many of which, as here, fail to hold up to scrutiny. While it would be foolish to deny that such a purpose features prominently in the thinking of many Americans, it would be equally foolish not to acknowledge that for many others religion is something else entirely. Religion is also about cultural identity.

When I am asked about my religion, I always reply that I am Jewish. I hardly think I can be accused of hiding my true beliefs, but the fact remains that I was born and raised in a Jewish home. That means something to me. It is unfortunate that so many people see these cultural differences as reasons for intolerance or even violence. Placed in their proper perspective, it is precisely these, ultimately trivial, differences that add color and texture to life.

Even as a child I was aware of this. My circle of acquaintances was neatly divided into my Jewish friends and my non-Jewish friends. Not, of course, because there was any really important difference between them. It was just that I had certain shared experiences with my Jewish friends (going to Hebrew school, celebrating Jewish holidays, putting up with well-meaning *goyim* unable to comprehend not celebrating Christmas) that I did not have with my non-Jewish friends.

I still get a chuckle when I think of how my non-Jewish friends responded to my Bar Mitzvah. The Saturday morning service was a miserable, three-hour affair that went from nine o'clock to noon. Most of my part of the service took place between ten and eleven. I patiently explained to my non-Jewish friends that it was perfectly appropriate to walk in just before ten and that no one would look at them funny for doing so. My Jewish friends, of course, did not need to be told this.

So what happened? On the big day I showed up at the synagogue around half-past eight. A good chunk of my non-Jewish friends were already hanging around, worried about being late. Having failed so completely to explain to them Jewish attitudes about punctuality, you can imagine what they thought when I told them the candy in the basket by the door was not for eating. Instead, it was to be thrown at me at a specific point in the service. I think they were trying to determine the precise species of barbarism in which they were about to participate.

Incidentally, my Jewish friends walked in, en masse, at 9:59.

Many years ago one of my Sunday school teachers told the class a story.[1] It was December, she recounted, and she was at the local mall with her daughter, then four. Seeing the other children sitting on Santa's lap, her daughter asked if she could do likewise. Figuring the child wasn't going to convert from a brief conversation with Santa, she said it was fine. After waiting on line for a while, her daughter climbed up on Santa's lap. Santa asked, "What do you want for Christmas, little girl?"

Her daughter looked at Santa and said, "I'm Jewish, I'm a Hanukkah person."

Santa laughed and said, "So am I."

Not long ago I saw a production of the musical *Fiddler on the Roof*. If you are unfamiliar with the story, it centers around a community of Jewish farmers in a small Russian village, set at the turn of the nineteenth to the twentieth century. The main character is Tevye, who, with his wife Golde, is raising five occasionally rebellious daughters amidst a gradually worsening political situation. In one scene Tevye and Golde have an interesting exchange. I should remind you that this sounds better when it's sung, but the words are powerful all by themselves:

> TEVYE: But my father and my mother said we'd learn to love each other, and now I'm asking, Golde, do you love me?
>
> GOLDE: I'm your wife!
>
> TEVYE: I know—but do you love me?
>
> GOLDE: Do I love him? For twenty-five years I've lived with him, fought with him, starved with him. Twenty-five years my bed is his, if that's not love, what is?
>
> TEVYE: Then you love me?
>
> GOLDE: I suppose I do.
>
> TEVYE: And I suppose I love you too.
>
> BOTH: It doesn't change a thing, but even so. After twenty-five years, it's nice to know.

Gets me every time. It reminds me of my maternal grandparents, who were themselves Central European Jews. I find it hard to imagine my grandfather asking something so mawkish and sentimental, but if he ever did that is precisely the sort of bemused, super-practical answer my grandmother would have given him. As I watched the show there was scene after scene where I felt I understood these people. Not because I know any nineteenth-century Russian Jews, but because Jews are Jews, and we recognize each other.

To put it simply, I love being Jewish.

I love that for all my mordant atheism I am not even one whit less Jewish than the most orthodox rabbi. There are no good Jews or bad Jews or lapsed Jews or anything Jews. There are just Jews, and that is all.

I love that a rabbi derives his authority not from any perceived personal relationship with God (an idea that Jews the world over regard as absurd and arrogant), but rather from his education and his years of study of all things Jewish. You respect rabbis on Jewish questions for the same reason you respect biologists on biology questions; they know more about it than you do. And if, for all that, you find yourself disagreeing with the rabbi, then it is perfectly acceptable to say so. Preferably with vehemence.

I love that Jews not only do not proselytize, but we positively discourage conversion. This attitude is not the result of any strong concern over whether the person really means it. Nor is it because we fear he might be converting just to please the in-laws. Not at all. It's because there is an implied world-weary question at the base of every Jew's existence that goes something like, "What do you want to be Jewish for?"

I love the part of the Passover seder where we express dismay that freedom from bondage came at the cost of so many Egyptian lives. We show this dismay symbolically by dipping our pinkies into our wine glasses and leaving ten drops of wine (one for each of the plagues) on our plate, thereby lessening our pleasure. This process is followed by the equally satisfying ritual of licking the excess wine off our fingers with a loud smack of the lips, as if to make clear that, let's be honest, we're not really sorry that those *farkakte* Egyptian taskmasters got what was coming to them. After all, those ten drops can be discreetly reabsorbed into the collective with a strategically placed dollop of *charoset*. Symbolism is good, but tasty wine is better!

I love that we celebrate our winter holiday not by killing a tree and decorating it with gaudy baubles, but by lighting a few candles, reciting two quick blessings (remembering to add the *sheheheyanu* on the first day), and eating chocolate coins. Far more civilized.

I love knowing that the proper condiment for a sandwich is not mayonnaise, a vile concoction whose sole legitimate purpose is found in good tuna salad, but mustard. Not just any mustard, mind you. Not the soulless neon-yellow glow-in-the-dark stuff or the vinegar with yellow food coloring or any of the other varieties of inedible glop so familiar from the condiment aisles of the nation's supermarkets. I mean a proper deli mustard. The kind with brown specks that comes in a small metal container whose lid flips up with gentle pressure from your thumb and has a small groove in it for the spoon. The kind that has a dish of sour pickles, pickled tomatoes, and cherry peppers next to it, so that the smell of salt, vinegar, and spices mixes seductively with the pile of fatty meat on the plate in front of you—the kind that also has a few half-sour pickles that you must never eat, unless you specifically want everyone to think you're a wimp. And I love the fact that every Jew reading this knows perfectly the emotions I am describing, while most of the non-Jews are wondering what I have against mayonnaise.

Actually, that reminds me of a story. A Christian friend of mine in graduate school (one of the same ones who saw Luis Pulau with me back in chapter 4) used to keep the raw materials for making sandwiches in the refrigerator in the student lounge. One day I noticed that he was putting mustard on his sandwich, and I commended him for his choice. "I didn't think Christians did that," I said. He smiled and replied, "Well, you guys aren't wrong about *everything*."

It's not all good news, of course. Do I believe that God made a covenant with my ancestors in which we agreed to live by certain implausible laws in return for the land of Israel? Of course not. That idea is *ridiculous*.

I have no urge to adhere to obscure dietary laws or to wrap myself up in phylacteries. It's just that, for all the dubious claims of the Torah, the fact remains that modern Judaism is almost exclusively focused on this world and not the next. We do not talk much about souls or the afterlife or our personal walk with God. Instead we talk about following the law, being part of the community, and getting non-Jews to leave us alone. I like that.

This focus on practical concerns was brought home to me in especially poignant fashion shortly after the terrorist attacks of September 11, 2001. I had been living in Kansas for just over a year. For the first time in my adult life, I was living in an environment largely devoid of Jews. In such circumstances even a hardcore secularist such as myself feels the call of the tribe, and I joined the local synagogue. I attended Friday-night services that week, and for obvious reasons the attacks were on everyone's mind. At one point the service was paused so that people could discuss their reactions.

The discussion went on for close to forty-five minutes. During that time the questions raised and comments made were all completely practical in nature: Interest in the best way of donating money to the relief and clean-up efforts. First-hand accounts from friends and family who lived or worked in New York. Concern for how the attacks could ultimately affect the state of Israel. Worry about possible retaliation against Muslim friends and acquaintances.

At no point did *anyone* bring up God, or implore people to pray, or raise theological questions about evil, or suggest that the victims of the attacks had, at least, gone to a better place. It made me proud to be a Jew.

There is no shortage of people who are culturally religious in the sense I have described here. Still, some may wonder whether this sort of thing is too unstable to serve any purpose. It is one thing to know abstractly that many people in the pews on Sunday do not really believe what the preacher is saying. It is quite another when most people are just going through the motions. In my own case, while I am happy to participate in Passover seders and Hanukkah candle lightings, and I have a *mezuzah* hanging next to my front door, the fact remains that Jewish practice plays a very small role in my life. Perhaps encouraging people to move toward a cultural, as opposed to theological, understanding of religion is just the first step toward the death of religion entirely.

Perhaps not. In his book *Society without God*, sociologist Phil Zuckerman points out that even in Sweden and Denmark, two of the least religious countries in the world, majorities still identify as Christians even while rejecting all of the major tenets of the faith. People still have their children baptized, and they continue to get married in churches. They do not protest the use of their tax dollars to support their nation's official church. Their general lack of supernatural beliefs has not led to the death of the church as an institution.

Were more of the world to shift away from theological religion toward cultural religion, many would still choose not to participate, but no one would be writing angry polemics about it. Sadly, we can only look forward to that far-off day. For now there are simply too many people who will be horrified by the suggestions I have made here. They will see my casual, mostly light-hearted discussion of religious differences and rituals as deeply disrespectful. They will not agree that anything of value remains when the dogmas and doctrines have been, if not removed, at least put in their proper, harmless, place. Some might even feel compelled to inform me about where my ecumenism will get me in the afterlife.

So be it. For now we can only acknowledge a spectrum of religious belief from atheism and purely cultural religion on one side to fundamentalism on the other. I know where I am on that spectrum. And I share with the fundamentalists the conviction that our society would be a far better place were more people to move to my way of thinking.

PART V

The Sixth International Conference on Creationism, Pittsburgh, Pennsylvania, August 2008

29

Building the Creation Model

Scientists mostly regard "creation science" as an oxymoron. We take it for granted that the "two-model" approach to science education, in which the creation model is cast as a legitimate scientific alternative to evolution, is just a sham invented as part of a failed attempt to inject creationism into the public schools. There is ample justification for that belief, in the form of creationism's manifest scientific errors and the frequently appalling behavior of its leading proponents.

Which is why—especially given the fundamentalist pyrotechnics I have witnessed at other conferences—I was surprised by the sixth annual International Conference on Creationism (ICC), held in Pittsburgh, Pennsylviania, in August 2008. The ICCs are creationism's attempt to hold a scientific research conference. The speakers in Pittsburgh, many of whom had legitimate credentials, approached their presentations with the utmost seriousness. The conference proceedings were a highly professional volume, with separate sections for earth science, stellar and planetary sciences, life sciences, foundations of science, and social science and the humanities.

To give you a sense of what high-level creation science looks like, here are a few paper titles that caught my eye in the proceedings:

- *Ocean Circulation Velocities over the Continents during Noah's Flood*
- *Distant High Energy Sources and the Cosmic Microwave Background in a Creation Day One Framework*
- *Radiohalos in the Sharp Granite, Lake District, England: Evidence That Removes Objections to Flood Geology*
- *Rapid and Early Post-Flood Mammalian Diversification Evidenced in the Green River Formation*
- *Structural Dynamic Stability of Noah's Ark*

These are just a few items, chosen specifically for their interesting combination of scientific and religious terminology.

It is easy to dismiss presenters of such papers as cranks, or to mock them for thinking that good science begins by declaring as sacrosanct certain dubious biblical propositions. For the moment, though, I would stress two other points.

FIGURE 29.1 Creationist children's literature on display at the conference.

First, we should have no doubt regarding the seriousness of the conference participants. We are not talking here about the professional creationists, the ones whose livelihood is spreading propaganda and corrupting school boards. We are talking instead about people who, so far as I can tell, are motivated by entirely the same considerations as mainstream scientists. They are trying to understand nature as best they can. That they begin from a premise most of us would regard as highly improbable has no relevance to that determination. Second, note that YECs have been far more successful than ID proponents at translating their ideas into scientific research. Since ID is vague to the point of vacuity, it is hard to imagine how ID folk could hold a conference of similar breadth or depth.

One of the organizers presented the following capsule history of the ICCs:

> So the first two ICCs, in 1986 and 1990, were devoted to the age of the Earth and the universe. . . . Then, in 1994 we started to have a different venue, if you will. We felt that what we need is to develop in a systematic and rigorous way the creation model. And there was really no organization per se, in the world, that was specifically geared to doing that. Developing the creation model of origins in a rigorous way with peer-reviewed papers. And so for the last four ICCs, 1994, 1998, and 2003, and now 2008, each conference has been devoted to that theme, developing and systematizing the creation model of origins.

I participated in the question periods after several of the talks, always to offer a contrary view. I would emphasize, as I have before, that at no point did anyone try to shut me down, boo me, or chase me away from the microphone. I could probably have participated more aggressively than I did without provoking anyone to anger. My suspicion is that most of the audience members were happy for the excitement.

This led to a funny moment on the third day of the conference. As I was preparing to leave a just-concluded presentation, two gentlemen with whom I had been conversing the day before walked in. There were maybe twenty other people milling around. With a booming voice and a huge smile, one of them pointed directly at me and called out, "I know *that* guy! He's a troublemaker." Most of the other folks had no idea what he was talking about, but looked in my direction anyway. I feigned offense and replied, "Me? Nah, I'm a pussycat." We both laughed.

One talk I found interesting was called "A Critique of the Precreation Chaos Gap Theory." The gap theory is a form of old-earth creationism I did not consider in chapter 26. It comes in various forms, all of which posit a long gap of time between two of the early verses of Genesis 1. In this way the integrity of the creation account can be preserved while allowing for an Earth of ancient age. For obvious reasons, the YECs do not approve. This talk stood in stark contrast to the standard caricature of creationist exegesis. Many is the time I have heard them accused of thinking that English is the original language of the biblical text or that the King James version is the only acceptable translation. By contrast, this speaker engaged in a minute analysis of the Hebrew text, and meticulously compared numerous translations, to come to some conclusions as to its meaning. I am not in a position to assess the merits of this work, so I pass no judgment on the correctness of his conclusions. I was nonetheless impressed by the level of seriousness that was brought to the task. I am well familiar with frothing, creationist propaganda. This was something different.

Let me close this chapter by mentioning the talk "A Review of the Search for Noah's Ark." I thought you might be curious about the status of that search, so here is the abstract:

> There have been many alleged sightings of the Ark and numerous attempts to find it, mainly on Mt. Ararat, but search attempts so far have been without success. In the light of history, geology, and archaeology, we need to consider that the Ark probably landed elsewhere, and that there may be little left of it.

30

Inevitable Humans?

We have previously discussed three of the main points of tension between evolution and Christianity: evolution at least potentially conflicts with the Bible, it exacerbates the problem of evil, and it seriously weakens the argument from design. A fourth point of tension featured prominently at the International Conference on Creationism, both in the formal presentations and in my conversations with other attendees. I am referring to the issue of whether evolution diminishes the role of humanity in God's creation.

To illustrate the problem, permit me a juxtaposition of quotations. The first is the opening sentence of the prologue to the Catechism of the Catholic Church: "God, infinitely perfect and blessed in himself, in a plan of sheer goodness freely created man to make him share in his own blessed life." The second comes from paleontologist Stephen Jay Gould. Referring to the Burgess Shale, a rock formation rich with fossils from the earliest stages of animal life, he writes, "Replay the tape a million times from a Burgess beginning, and I doubt that anything like *Homo sapiens* would ever evolve again" (Gould 1989, 289). It is hard to see how the Church and Gould could both be right. Is humanity the reason for creation, or are we just one of many improbable end results of evolution?

The fear that evolution lowers humanity's status is front and center in creationist thought, but perhaps there is a way out. We could hypothesize that Gould was wrong, and that humans were actually an evolutionary inevitability. If that is true, then we might plausibly see evolution as God's mechanism of creation. And since this issue is a major stumbling block for Christians trying to make their peace with evolution, we should inquire as to what can be said on behalf of this idea.

Whatever conclusion we may form regarding the correctness of Gould's view, we can certainly say his opinion is widely shared among scientists.[1] Writing in 1964, paleontologist George Gaylord Simpson spoke for many:

> The assumption, so freely made by astronomers, physicists and some biochemists, that once life gets started anywhere, humanoids will eventually and inevitably appear is plainly false. The factors that have determined the appearance of

> man have been so extremely special, so very long continued, so incredibly intricate that I have hardly been able to hint at them here. Indeed they are far from all being known, and everything we learn seems to make them even more appallingly unique. (Simpson 1964, 774)

The basis for this conclusion is clear. Natural selection knows no foresight. It understands only short-term success in passing genes into the next generation. The genetic variations on which it acts are themselves random with respect to the needs of the organism. Moreover, the fitness of a given variation often depends on the environment, so that mutations advantageous in one environment might be harmful in another. These environments themselves change in unpredictable ways, introducing yet another realm for chance. What directing principle can be found in any of this?

Nor should we overlook the role of mass-extinction events. Several such events are recorded in the fossil record, including the one that killed off the dinosaurs (generally thought to be an asteroid impact). Had this not occurred it is unclear that mammals would ever have become prominent. Gould writes:

> If mammals had arisen late and helped to drive dinosaurs to their doom, then we could legitimately propose a scenario of expected progress. But dinosaurs remained dominant and probably became extinct only as a quirky result of the most unpredictable of all events—a mass dying triggered by extraterrestrial impact. If dinosaurs had not died in this event, they would probably still dominate the domain of large-bodied vertebrates, as they had for so long with such conspicuous success, and mammals would still be small creatures in the interstices of their world. . . . Since dinosaurs were not moving toward markedly larger brains, and since such a prospect may lie outside the capabilities of reptilian design, we must assume that consciousness would not have evolved on our planet if a cosmic catastrophe had not claimed the dinosaurs as victims. (Gould 1989, 318)

What can defenders of human inevitability offer against this?

We shall begin with three possibilities that, I believe, can be dismissed fairly quickly.

The first is that of complete causal determinism. It is possible that everything that has happened since the moment after the Big Bang has been foreordained by the laws of physics. A sufficiently intelligent being in possession of perfect information could have predicted everything that has happened since the moment of creation. What we perceive as chance and contingency, whether in biological evolution or in modern quantum mechanics, is just a reflection of our own ignorance.

This possibility cannot be ruled out, but it hardly seems like a worldview designed for spiritual comfort. If you find it appealing you are welcome to it, but I will not consider it further here.

Another possibility is that God is subtly involved in the evolutionary process, in ways that are beyond our ability to detect. Physicist Robert John Russell defends a version of this view, suggesting that God avails Himself of quantum indeterminacy to influence the mutations that occur (Russell 2003). In this way He can ensure that evolution produces human beings without engaging in flamboyant violations of His own natural laws. This proposal has the virtue of being empirically irrelevant, meaning there is no danger that science could actually disprove it.

I would think, though, that it creates more theological problems than it solves. With God once more directly involved in the evolutionary process, the problem of evil returns. If God intervenes to ensure that evolution produces desirable things, how then is He not to blame for the undesirable things? We are also left wondering why God would create in so bizarre a fashion. Others take it as a sign of God's greatness that He can allow the world to create itself through His laws (Haarsma 2003), but here comes Russell to hypothesize that God is endlessly tinkering and fiddling. Also, Russell's hypothesis, no less than the idea that fossils were made to appear old to test our faith, places God in the role of deceiver. Our conclusions about genetic randomness and quantum indeterminacy rest on a mass of data, and they are central to our current best understanding of the universe. That this is an illusion that covers God's actions is as offensive to the scientific sensibility as outright violations of His laws.[2]

Our third possibility comes from philosopher Michael Ruse. He agrees with Gould that the evolution of humans was a highly improbable event, but notes that this conclusion holds only if we are permitted a single play of the tape. We might hypothesize, therefore, that God created a large number of universes, confident that humanity was effectively guaranteed to appear in at least one of them (Ruse 2010, 221–222). This is logically possible, but it forces us to face anew the question of why God creates through a process and does not simply go straight to the point. It was hard enough to understand why God set the Darwinian wheels in motion once. Now we are to imagine He did so countless times, just so humans could appear at least once. This seems terribly roundabout. We are back to the problem we considered in chapter 24, trying to find a reason for supposing that humans can only be created by Darwinian means.

There is a great irony, though, in Ruse's suggestion. We briefly encountered the idea of multiple universes in chapter 19, where we suggested them as part of a reply to the "fine-tuning" argument. Their effectiveness in this regard has led many theologians to scorn the whole idea. Thus, John Haught writes:

> Of course, if you are truly addicted to the idea that our life-bearing universe is a purely random, undirected, and unintelligible occurrence, and that life within it *must* in no sense be the product of divine intelligence and wisdom, you may then imaginatively conjure up an endless series or proliferation of other "universes," so as to increase the probability that randomness rules. (Haught 2000, 35–36)

Physicist-turned-theologian John Polkinghorne offers similar thoughts:

> The multiverse theory in its more extreme forms is the idea that there are these vast portfolios of different universes, disconnected from ours, unobservable by us. It's a metaphysical guess. It has mostly been popular and mostly been invented in order to explain away the fine-tuning of our particular universe. (Fitzgerald, 2008, 33)

Somehow I don't think theologians will be happy with Ruse's suggestion.

I would add, though, that the remarks of Haught and Polkinghorne amount to little more than obscurantism. That the reality of the multiverse is strongly suggested by the mathematics underlying much of modern physics is no longer in doubt. Nowadays the debate has moved on to the precise sort of multiverse we inhabit (Greene, 2011). It is nonsense to suggest that multiple universes were hypothesized into existence to deal with "fine-tuning."

At any rate, if these possibilities are the best we can do, then we should not be surprised that so many Christians have a problem with evolution. Let us investigate more promising approaches.

In asking if humans would evolve a second time, we are not referring to the precise physical form of modern *Homo sapiens*. We are asking instead about humanlike intelligence. We assume that God does not care about our physical appearance, but only about our ability to enter into a relationship with Him. With that in mind, biologist Kenneth Miller demurs from Gould's view:

> Turning our attention to the special case of our own species, we can be fairly confident, just as Gould tells us, that our peculiar natural history would not repeat, and that self-awareness would not emerge from the primates. Indeed, we would have no reason to suppose that primates, mammals, or even vertebrates would emerge in a second running of the tape. But as life reexplored adaptive space, could we be certain that our niche would not be occupied? I would argue that we could be almost certain that it would be—that eventually evolution would produce an intelligent, self-aware, reflective creature endowed with a nervous system large enough to solve the very same questions that we have, and capable of discovering the very process that produced it, the process of evolution. (Miller 2008, 152)

The concept of a "niche" is difficult to define precisely, but roughly it refers to the sorts of things an animal does to make a living. An animal exists within a certain environment, eats certain things and gets eaten by others, competes for resources with other species—all of these factors combined comprise its niche. "Adaptive space" is then an abstract way of pondering all the different niches in nature. Miller's claim is that evolution will inevitably fill the same niches over and over again in each "replay of the tape."

He bases this conclusion on the phenomenon of evolutionary convergence, in which the same adaptation evolves multiple times in different lineages. Sharks and dolphins parted ways long ago, but both arrived at the same streamlined shape so appropriate to life in the water. This is one example of a common phenomenon. If different lineages routinely converge on the same adaptive solutions to their problems, then that is strong evidence that evolutionary outcomes are less contingent than Gould suggests.

Paleontologist Simon Conway Morris has been an especially passionate advocate of this argument. He describes the evolution of human intelligence as a "near-inevitability" (Conway Morris 2003, xii) and writes, following a discussion of convergent evolution in the sensory organs of various species:

> Could the story of sensory perception be one clue that, given time, evolution will inevitably lead not only to the emergence of such properties as intelligence, but also to other complexities, such as, say, agriculture and culture, that we tend to regard as the prerogative of the human? We may be unique, but paradoxically those properties that define our uniqueness can still be inherent in the evolutionary process. In other words, if we humans had not evolved then something more-or-less identical would have emerged sooner or later. (ibid., 196).

In what follows I will refer to the argument made by Miller and Conway Morris as "the argument from convergence."[3] In this section we consider three basic objections that should make us suspicious of the argument. Starting in the next section, we shall discuss things in more detail.

The argument from convergence, especially as presented by Conway Morris, uses the repeated evolution of a structure (eyes, for example) as evidence that its appearance is likely to recur on a replay of the tape. Philosopher Elliott Sober notes a problem with this:

> You can't show that an event was inevitable or highly probable just by pointing out that it has happened many times. To estimate the probability of the camera eye's evolving, you need to know how many times it evolved and how many times it did not. Conway Morris never describes how often convergences failed to occur. (Sober 2003)

We should note, as an aside to this objection, that other scientists have strongly disputed the extent to which convergence occurs (Gould 1998; Prothero 2003).

A second problem is that "evolutionary convergence" does not have a precise definition. There is no standard for how similar two adaptations must be to qualify as a convergence. Philosopher Kim Sterelny raises this issue forcefully:

> Under what conditions should we say that the same trait has evolved again? The problem is that there are always differences in detail, and sometimes more than detail, when similar traits evolve independently of one another. . . .
>
> One problem is that traits can be described in fine- and coarse-grained ways. Choose coarse-grained descriptions, and we multiply the number of

times "the same trait" has evolved independently. Choose fine-grained descriptions, and we shrink the empire of convergence. (Sterelny 2005, 592)

Sterelny then points to several cases in which Conway Morris's claimed convergences require very coarse-grained descriptions indeed.

The next point is that the adaptation in which we are most interested is human intelligence, and it only evolved once. If multiple recurrences of an adaptation are treated as evidence for inevitability, then why is the uniqueness of human intelligence not evidence for its improbability? Anything more than rudimentary intelligence is quite rare in nature, and we can find nothing among animals to rival the human brain. Biologist David Zeigler writes:

> Having the brain power to recognize and remember individuals in your social group, along with their typical behavior and personalities, has indeed evolved in dolphins, elephants, and some primates (traits otherwise extremely rare in the animal kingdom). These achievements, however, are not on the same level as the singular accomplishments of complex symbolic language, high-level reasoning, and creativity that humans alone have achieved. (Ziegler 2008; see also Coyne 2009a)

None of these points flatly refute the argument from convergence, but they do show that its supporters have more work to do.

In discussing the forms of organisms, it is helpful to use the concept of "morphospace." By this we mean the collection of all possible animal forms, arranged in such a way that animals with similar forms are located close together. Central to the present debate is the undeniable fact that actual animals are not distributed evenly through morphospace. Instead there are clumps, as disparate lineages tend to converge on similar forms. This clumping requires explanation.

Defenders of the argument from convergence explain the clumping primarily by reference to forces external to the organism. Natural selection repeatedly discovers the same optimal solutions to the same problems of functional design. For example, both sharks and dolphins must move gracefully through water, so natural selection ultimately guides them to the same streamlined body shape. Defenders of contingency emphasize instead the role of internal limitations. Since major reorganizations of body plans are not generally possible, what evolves in the future is severely constrained by what has happened already in the past (Powell 2007).

This leads to an important distinction between "parallel evolution," on the one hand, and "convergent evolution," on the other. Once again the line is often hard to draw sharply, but the general idea is clear. If two closely related lineages respond to the same environmental stresses by evolving in the same direction, we call it parallel evolution. When widely separated lineages perform the same feat we call it convergence. In the context of assessing human inevitability, convergence is more impressive evidence than parallelism.

This poses a problem, since Sterelny (2005) notes that many of Conway Morris's alleged examples of convergences are really better classified as parallelisms. The problem worsens when we consider the phenomenon of "deep homology," which refers to fundamental developmental mechanisms that are shared across even widely separated lineages. The pervasiveness of this phenomenon suggests that all evolution is parallel to some degree. Paleontologist Donald Prothero explains the significance of this:

> Once groups of organisms are established and develop a body plan and set of niches, biological constraints are such that convergence and parallelism can be expected. But the issue of who gets this head start may be more a matter of luck and contingency that has nothing to do with adaptation. (Prothero 2003)

To picture this, imagine a large table with a collection of cones arranged upon it. The wide mouths of the cones are pointed upward and arrayed closely together, while the narrow points rest on the table itself. Now toss a ping-pong ball onto the table. Without a detailed knowledge of the ball's trajectory we cannot predict in which cone it will land. Once the ball finds a cone, however, we can be certain that it will roll down to the bottom. There is a role for chance and a role for inevitability. Since small changes in the initial conditions can cause the ball to land in a different cone, and since the points of the cones are widely separated, we see that chance fluctuations at the start of the process can lead to big differences at the end.[4]

This leads to our final notion, that of "conditional inevitability" (Sterelny 2005). The ping-pong ball inevitably finds the bottom of its cone, but that is conditional on having landed in that cone to begin with. Likewise for evolutionary trajectories. For any given adaptation, we can imagine a statement of the form, "If evolution reaches stage *X*, then it is close to inevitable that the adaptation will appear." It is commonly thought that eyes, for example, have evolved many times independently. This provides some basis for thinking that in the case of eyes, "stage *X*" is something that occurs with high probability early in the evolutionary process. The question is whether the comparable stage for humanlike intelligence is also reached with high probability once the evolutionary process gets started.

An empirical test to determine the relative roles of chance and inevitability in evolution would be nice. There have been some attempts in that direction (Losos et al. 1998; Travisano et al. 1995), but it is doubtful that they can really address the sorts of large-scale contingency Gould had in mind in referring to what would happen were we to replay the tape from a Brugess beginning (Beatty 2006; Turner 2011). We might hope that the fossil record could shed some light, but Raup (1989) provides strong reasons for pessimism.

Still, the fossil record does provide us with a rich trove of hominid fossils, and I would think there is one salient fact that should make us suspicious of claims of human inevitability. The Neanderthals dominated Europe for some two hundred thousand years, with no competition from other hominid species, but there is nothing to suggest they were moving toward the intelligence of modern humans (Shermer 2008; Gould 1989, 319–320). Paleontologist Richard Klein writes:

> It is not difficult to understand why the Neanderthals failed to survive. The archaeological record shows that in virtually every detectable aspect—artifacts, site modification, ability to adapt to extreme environments, subsistence, and so forth—the Neanderthals lagged their modern successors, and their more primitive behavior limited their ability to compete for game and other shared resources. To judge from their distinctive morphology and their genetic difference from living humans, their more primitive behavior may have been rooted in their biological makeup.[5] (Klein 2009, 586)

It would seem, then, that even evolving an upright biped with opposable thumbs is not yet sufficient to ensure the appearance of humanlike intelligence. Surely that should give us pause when considering the possibility of human inevitability.

So what is the conclusion? Alas, we have reached a familiar impasse. We lack the data to provide a definitive resolution, a situation that is unlikely to change any time soon. I would argue, though, that such data as we have militate against the notion that humanlike creatures would evolve a second time. At the very least we can say that phrases like "almost certain" and "near-inevitability" have no place in this discussion. To the extent that traditional faith requires that humans play a central role in the creation, it is perfectly reasonable for Christians to be concerned about this.

31

Unpleasantness

An interesting exchange took place during the Q&A of a talk entitled "Georgia Public School Board Members' Beliefs Concerning the Inclusion of Creationism in the Science Curriculum." The speaker was Kathie Morgan of Liberty University. The talk itself was unremarkable. Its premise involved methods for bringing creationism into the classroom, in the form of supplementary materials beyond what state requirements mandate, that do not run afoul of any Supreme Court rulings. Morgan and her colleagues then investigated the relationship between the personal beliefs of school board members and their receptivity to the idea of teaching creationism.

For me, the only part that provoked a raised eyebrow was when Morgan opened with a prayer, which, were it to happen at a normal science conference, would be considered eccentric. I remained in my seat as people approached the microphones for the question period. After a few fawning and perfunctory queries, a woman identifying herself as affiliated with the Institute for Creation Research Graduate School approached the microphone. She protested the bias against them in the educational establishment and encouraged parents to strengthen the faith of their children as protection against evolutionary propaganda. I was only half-listening, since this was all standard boilerplate. But when she said, "It's almost like Hitler's time," I quickly snapped to attention. She continued:

> It's almost like we are being so indoctrinated, and our teachers are being taught, in our state schools, the answers to give when students ask questions. And furthermore it's in our state objectives we are supposed to be able to teach both the weaknesses and strengths of evolution, but there's hardly anyone who knows the weaknesses of evolution because nobody teaches it. And we need to get people together and have a plan in place for the education system, so we can move forward. The ACLU can put out all these little fires all over the place just like in Georgia. They see something come up and they send a bunch of their Gestapo people there to thwart whatever is going on. And we need to have some sort of meeting of educators so we can start doing something about it.[1]

Somehow the situation she described did not, for me at least, summon forth images of Hitler. Thinking a reply was called for, I got in line at the microphone. After waiting patiently for a few other questioners to finish, (one of whom was eager to inform us that Darwinists know they cannot refute the evidence for a young Earth and rely on repressive tactics to cover up this deficiency), I said this:

> I don't really have a question but I would like to make a comment to the speaker a few questioners ago. I would request that you not make such casual comparisons to Hitler. When you're forced to wear yellow stars as a prelude to getting rounded up into train cars and the rest of it, then you can compare your situation to Hitler. And don't liken the ACLU to the Gestapo. These are people who are filing lawsuits in a legal proceeding to protest what they believe are violations of civil rights, and these lawsuits then get adjudicated through a lawful process. That ain't the Gestapo. So I would ask that you tone it down a little bit.

Feeling good about myself, I returned to my seat. The woman, to her credit, looked somewhat abashed. Less impressed was an older gentleman who approached the microphone to reply. It was Jerry Bergman, a prominent creationist speaker and writer. He said,

> Let me respond to that. I just finished a book called *Slaughter of the Dissidents*, with the subtitle *The Shocking Truth About Killing the Careers of Darwin Doubters*. It's three volumes, fifteen hundred pages. I interviewed hundreds of people whose careers were ended, they ended up in divorce and suicide, quite a few suicides. There's quite a bit of physical violence, people who have been beaten up. And it's true, your point is well-taken. . . .

From my seat I called out, with maximum disgust, "You're comparing that to six million people who lost their lives." Bergman continued:

> That's true, we're not being put into concentration camps. But I know people who have not worked in twenty years. I know a person with two doctoral degrees who has been unable to find work in fifteen years. It's depressing, doing this. It's very depressing, my wife won't read it, she proofs most everything I do and she will not read this because she says it's too depressing as to what's going on. It really is, in many ways we're faced with enormous opposition and I see it as getting worse. I hope not, I hope I'm wrong. But pick up my book. It should be at the printer this week and if you want information about it I'll be glad to give it to you.
>
> But it's really a major problem, and the Darwinists are really getting vicious, they really are. We are, I would say, following the history of Nazi Germany. They went through four periods of persecution against the Jews specifically, we're in the second period in this country. And it's true we have two more to go and I hope it doesn't go there, but it's edging up there especially when you realize what happens in the lives of these people. And it's a huge waste of resources. What bothers me most is how vicious the Darwinists

> are. Really, really vicious people. I'd love to argue about this and I don't like to talk so much with people who agree with me. I'd rather talk to people who disagree with me. It's far more invigorating. And I've found you can't dialogue with Darwinists by and large. You just can't do it. It's a barrage of name-calling. Up on the platform I have several times literally had people come up on the stage and try to pull me off the stage. I've had people threaten my work. The college is getting tired of people calling, the college where I teach at, of trying to get me fired. They're tired of it, it happens so often. Fortunately they've defended me. But many schools, like Gonzales, one reason he lost his career was because the college got tired of people calling up and saying you've got to fire this guy because he's an ID supporter. He's a theistic evolutionist, by the way. They can't even deal with theistic evolutionists they only can accept more and more atheists. So I'm glad we have Liberty University around, we've got a few schools, so read what's going on and you won't have such a benign opinion about what's happening. It's really frightening.

There is, of course, much to object to in this statement. Creationists are not exactly amateurs in the vicious name-calling department, and that hardly seemed like an apt comparison to Hitler's Germany regardless. Creationist urban legends about "Darwin Doubters" having their careers summarily destroyed almost never include all of the relevant facts.[2] I was tempted to get up again to reply, but the moderator stepped in to inform us that time was fast running out and that only the people already standing would have a chance to ask their questions. Probably for the best, since I was angry enough to say something regrettable.

After the talk quite a few people thanked me for my comments. On several occasions people later stopped me in the hall to tell me they agreed with the point I had raised. One complained to me pointedly of the "non-Christian" manner in which some of his fellow creationists comported themselves.

Their approbation was encouraging. From my own side I could think of numerous examples of evolutionists unleashing a level of invective towards creationists that I found off-putting and uncalled for. The whole incident reminded me anew that in many cases the decent and generous people in the audiences are often not well represented by their leaders.

Continuing with the theme of outbreaks of unpleasantness into the otherwise staid proceedings, let me mention something that arose in a talk given by Phil Pantana. According to his website, Pantana once ran for president under the slogan, "One Nation, One God, One Law." His talk was a standard melange of creationist talking points, but the part I want to tell you about is this: Pantana said, "He's [Charles Darwin] going to be one hundred fifty years old, or, is it his book [*The Origin of Species*], when his book came out? It was a racist book, it talked about special, favored races." This was just an aside, tossed off while fumbling to find the correct slide in his presentation.

This particular meme is ubiquitous in creationist discourse. Ken Ham and A. Charles Ware elaborate on the idea in their book *Darwin's Plantation: Evolution's Racist Roots*:

> Although racism did not begin with Darwinism, Darwin did more than any person to popularize it. After Darwin "proved" that all humans descended from apes, it was natural to conclude that some races had descended further than others. In his opinion, some races (namely the white ones) have left the others far beyond, while other races (pygmies especially) have hardly matured at all. The subtitle of Darwin's classic 1859 book, *The Origin of the Species*, was *The Preservation of Favoured Races in the Struggle for Life*. The book dealt with the evolution of animals in general, and his later book, *The Descent of Man*, applied his theory to humans.[3] (ibid., 22)

I think it is safe to say that these are not the words of scholars trying to understand the history and causes of racism. Sometimes it is best just to let insanity speak for itself, but I will nonetheless offer a few counterpoints.

There is something especially grotesque in using the image of a plantation with regard to Darwin and evolution. Darwin himself was staunchly antislavery, a stance, incidentally, which put him firmly at odds with many of the clergy of his time. His anti-slavery views played a strong role in his formulation of evolutionary theory. Historians Adrian Desmond and James Moore write:

> It was Darwin's starting point too, his abhorrence of racial servitude and brutality, his hatred of the slaver's desire, as he jotted, to 'make the black man [an] other kind', sub-human, a beast to be chained. Roots were what Darwin's human project was all about. And to understand why he started thinking about roots-the origin-of black and white races, we have to appreciate his moral anchorage in the noontide of the British anti-slavery movement. It is the key to explain why such a gentleman of wealth and standing should risk all to develop his bestial 'monkey-man' image of our ancestry in the first place. (Desmond and Moore 2009, xvii)

As Ham and Ware note, *The Origin of Species* did not address human evolution at all. That should make us suspect that the reference to "favoured races" in the subtitle meant something other than human races. In fact, the word "race" as used by Darwin is roughly equivalent to the term "subspecies" as used today. He was referring to small populations of animals, and they were "favoured" only in the sense of possessing some anatomical variation permitting them to outcompete their rivals.

I would like very much to engage in polite conversation with creationists, and as I suggested in the last section I have met many with whom that is possible. Sadly, though, as long as their leaders insist on producing this sort of ignorant venom such conversations will have to remain the exception.

32

Conversations in Bookstores

As always at conferences, the ICC featured a large bookstore. According to the organizers, it was the largest collection of creationist materials ever assembled.

Those who have studied science at a high level bring with them certain expectations regarding books in their subject area. Long experience grappling with subtle ideas leads to the recognition that serious scientific writing will have a certain density. You look for careful presentations of arguments and meticulous attention to definitions. You expect a copious bibliography. This is especially true in books purporting to offer novel approaches to well-established disciplines.

After long periods immersed in the real thing, you develop a nose for counterfeits. A book alleging that entire branches of science have been on a disastrous path for decades will be an object of suspicion for that reason alone. When that bold claim is defended entirely with arguments readily presented to a high-school student, the suspicion deepens. That these arguments require only a few paragraphs for a complete presentation is a further bad sign. And if you then note a smug, arrogant tone, as though the writer is the sole clear-thinker in a world of blinkered and confused scientists, you can, with great confidence, conclude that the book delivers far less than it promises.

As the ICC illustrates, there are people who attempt to undertake serious scientific research from a creationist perspective. Sometimes they write books and articles describing the fruits of their labors, but the audience for that sort of thing is quite small. People interested in high-level scientific prose typically have a background sufficient to make them aware of creationism's many flaws.

The offerings in the bookstore were mostly of a different sort altogether. They were works of such unabashed advocacy, or propaganda if you are cynical, that they were hardly works of science at all.

Late in the afternoon on the second day of the ICC, a young man approached me in the bookstore. We talked for roughly twenty minutes.

After some chitchat, I asked him directly why he was a creationist. He told me that he came to Christ when he was twelve, but gave no thought to the evolution question until several years later. He knew that many Christians accepted evolution, and at that time he did not really care one way or the other about it. Then he heard a sermon at his church discussing the issue from a creationist perspective. His interest piqued, he began reading up on both sides of the subject. He concluded that the creationists had the better arguments.

I asked him who he had read from the evolution side. He hesitated, surprised by the question. After some fumbling he only managed to name Stephen Jay Gould, which was suspicious, since Gould was so famous that he is often the only pro-evolution writer creationists can name.

We batted around the issues for a while, with neither of us budging from our starting points. With some exasperation I eventually said something like:

> The fossil record is an obvious place to look for evidence of evolution. If it were really the embarrassment you say it is, how do you explain that virtually every paleontologist in the world is an evolutionist? And if geneticists really could not explain how genetic information can grow over time, which is one of the fundamental questions any theory of evolution has to explain, why would evolution be accepted so overwhelmingly among them? And are you really suggesting that generations of scientists have simply overlooked an elementary contradiction between evolution and the second law of thermodynamics?

He did not miss a step. He told me that sin is a powerful force. It has systematically colored the way generations of scientists view the data.

But it did not color his own view of the data?

No, because he had the Bible to keep him anchored.

But what about all of the scientists who are Christians? Or all the Christians who accept evolution? Did they not also have the Bible?

They compromised the Bible's clear teaching. They placed the world's wisdom ahead of God's word.

But they are sincere believers. Many of them believe specifically that they are honoring God by carrying out research in evolutionary biology.

It's very sad they have been so led astray. Sin is a powerful force.

But why was evolution so thoroughly rejected prior to Darwin? Scientists like Lamarck became targets of scorn for suggesting the notion. If scientists were in thrall to sin, they should have jumped at evolution at the first opportunity. How do you explain that?

The way to apostasy was being paved by geologists like James Hutton and Charles Lyell and their theories regarding the antiquity of the Earth. Darwin arrived at the right time to take advantage of that.

But Hutton and Lyell were themselves Christians. They certainly did not have an anti-God agenda.

Sin is a powerful force.

As it happens, that was not the only interesting conversation I have ever had in a Christian bookstore.

I previously mentioned the three years I spent at Kansas State University. There was a Christian bookstore located near campus, and I occasionally went in to browse. This led to a number of noteworthy moments. For example, there was the time I asked the person behind the counter why their extensive evolution/creation section did not include any of the books arguing that evolution and Christianity were compatible. She laughed.

Then there was the time my browsing was interrupted by an excited voice from behind me. It said, “Hi, Dr. Rosenhouse!” I turned around and saw one of my students. She was working at the store that day. She looked happy to see me, which was surprising, since she had never seemed especially joyful to be in my class. We spent a few minutes chatting, during which time it became clear she simply assumed I shared the religious viewpoint of the store. Lacking the heart to tell her I was only there in the spirit of opposition research, I carefully avoided disabusing her of the notion. The conversation was a bit awkward from my side, but everything proceeded pleasantly enough. Perhaps it was my imagination, but she seemed considerably more engaged in class after that chance meeting. She certainly started coming to my office hours more frequently than before—that was not difficult, since she had never previously come to my office hours.

There is a basic sociological fact that can pose a challenge for those of us teaching math and science in culturally conservative parts of the country. As a group, scientists are far less religious than the public at large. Of the many studies that establish that basic conclusion, an especially thorough and recent one was carried out by sociologist Elaine Howard Ecklund (2010). She compiled data based on surveys of 1,700 scientists at ten elite universities. Her main findings are shown in tables 32.1 and 32.2.

Notice that 64 percent of scientists are atheists or agnostics; add the 8 percent who believe in a higher power that is not God, and we arrive at 72 percent that hold nontheistic beliefs. Moreover, more than one-half of scientist respondents claim no religious affiliation, compared to just 16 percent of the population. This is especially

TABLE 32.1

Scientists’ Beliefs about God Compared to All Americans

Belief about God	Scientists (%)	U.S. Population (%)
I do not believe in God	34	2
I do not know if there is a God, and there is no way to find out	30	4
I believe in a higher power but it is not God	8	10
I believe in God sometimes	5	4
I have some doubts, but I believe in God	14	17
I have no doubts about God’s existence	9	63

Source: Ecklund (2010)

TABLE 32.2

Religious Affiliations of Scientists, Compared to All Americans

Affiliation	Scientists (%)	U. S. Population (%)
Evangelical Protestant	2	28
Mainline Protestant	14	13
Black Protestant	0.2	8
Catholic	9	27
Jewish	16	2
Other	7	6
None	53	16

Source: Ecklund (2010)

striking, since many nonbelievers claim a traditional affiliation simply as a matter of cultural identity.

From the other side, evangelical Christianity is all but nonexistent among scientists, but represents more than one-fourth of the population generally. These findings are broadly typical of the studies that have been carried out. Among academic scientists, nonbelief and theologically liberal forms of traditional religions predominate.

Of course, it is easy to exaggerate. Traditional belief is a minority view, but it is sufficiently represented to show that many people manage to reconcile science with religion. I also will not wade into the difficult question of why success in science is so strongly correlated with nonbelief; that the correlation exists is sufficient for my purposes. All of that noted, the data is sufficiently stark to dampen the more enthusiastic pronouncements regarding science/religion compatibility.

33

Is the Earth at the Center of the Universe?

Also interesting at the ICC was a presentation called "Evidence for an Earth-Centered Universe." It was delivered by Mark Matthews, who had no institutional affiliation. His primary line of evidence involved "the fingers of God," which is the somewhat flamboyant name given by astronomers to a well-known phenomenon.[1] When they make charts of the locations of galaxies, they routinely find that a surprisingly large number of them fall on straight lines, as seen from the point of view of the observer. This is typically explained as an optical illusion arising ultimately from the expansion of the universe. All galaxies are receding from one another, which causes a redshift in the light we see when looking through our telescopes. As a consequence, the charts that we make of the locations of galaxies are based not on direct observations, but on inferences drawn from the magnitudes of the redshifts. Unfortunately, the large velocities and gravitational forces associated with clusters of galaxies can cause an alteration in the usual relationship between redshift and distance. The mathematics becomes very complicated, but for us it suffices to note that the "fingers" can be explained with entirely conventional physics. An observer in a galaxy far removed from ours would see precisely the same thing.

Matthews rejects that explanation. He sees the fingers as evidence that Earth occupies a special location in the universe. Just as an observer sitting at the center of a wheel would see something very different from an observer located along a spoke, so too does the universe look fundamentally different to an observer on Earth than it would to an observer elsewhere in the universe.

Here is an interesting quotation from Matthews's paper:

> Because the idea that earth holds a special place in the universe is anathema to modern cosmologists, a different explanation for the fingers of God had to be found. J. C. Jackson appears to be the first astronomer to discuss the fingers of God. Observe how Jackson breaks the news to the astronomical community that the earth's nonspecial position in the universe might be threatened:
>
> > The galaxies appear to fall into long chains or cigar-shaped configurations, all pointing at the earth. Unless one is prepared to assign to the earth a very

> special place in the universe, one must conclude that D is not a good distance indicator, and that in reality the galaxies exist in roughly spherical configurations whose internal velocity dispersions are several times that which would be observed if these systems were expanding with the universe.
>
> In the above quote, notice how Jackson immediately offers the astronomical community a way to avoid the obvious indication that earth occupies a special position in the universe. (Matthews 2008, 232–233)

This sort of snideness and casual motive-questioning would be deemed inappropriate in mainstream scientific journals, but in creationist writing it is *de rigueur.*

That Earth does not occupy a privileged place in the universe is a conclusion drawn from piles of empirical data. It is not an unquestioned premise put forth to avoid displeasing theological truths. The choice Jackson gave his readers involved discarding virtually every aspect of modern cosmology, on the one hand, or making a small adjustment in one esoteric area, on the other. That hardly seems like evidence of anti-religious bias.

During the question period, I went to the microphone to challenge the notion that the apparent lining-up of certain galaxies suggested that Earth occupied a special location in the universe. I said in part:

> Regarding these fingers of God, I think you're reading meaning into things that don't really have any meaning. . . . But here again an astronomer in a different galaxy could do precisely the same argument. And I think this is just exactly the same kind of reasoning that led people to think there was something significant in constellations—"Oh, that group of stars looks a little like a bear and that one looks like a big dipper"—but there wasn't really a pattern there, they were reading into that. And I think that's all you're doing.

The audience was not amused by my remarks. Calling out from their seats, various people told me to look again at the diagram. Quite a few of the galaxies really were lined up! Matthews joined the party by pointing to the fingers and saying that you can see them, plain as day. They aren't an illusion. I replied that constellations aren't an illusion either. There was a lot of crosstalk, with nothing helpful coming as a result.

Then a funny thing happened.

We had been asked to state our name before asking our question. Matthews, frustrated by his inability to convince me, now said: "Jason is a well-known guy on the Internet, he's working against us. Even secular cosmologists acknowledge these features and acknowledge that they are pointing towards Earth."

On the substance, that cosmologists acknowledge that the fingers point toward Earth is unsurprising. It is part of their understanding of the phenomenon that it causes the relative positions of the galaxies to seem elongated relative to the observer, standing on Earth in this case. But that is not what I was thinking at that particular moment. You see, I had not realized that I was a "well-known guy." Genuinely surprised, I said the only thing that came to my mind: "You've heard of me?"

That might have been the end of things, but two questions later a man approached the microphone and said this:

> I agree with you [pointing at me]. The fingers of God effect has been around a long time. This is actually quite an old map. This is a redshift space map that is being converted to real space using some Hubble parameter. So in actual fact you cannot say that's real space. It's not. It's redshift space. . . . And the finger of God effect that you can see on that line there, that occurs all over the whole sky, in clusters. And it's very easy to understand. If you have a spherical cluster of galaxies, and you add a dispersion velocity to that cluster, that simply means that cluster is in dynamic equilibrium.
>
> So any individual galaxy within the cluster has an orbit around its mutual center of gravity. So the cluster itself is in expansion with the Hubble flow, the center of mass of the cluster has a genuine cosmological redshift, that any individual component in the cluster has a random velocity component with respect to the center of mass of the cluster. Redshift only allows you to see the radial component of a real space component. . . . In other words, the velocity component of any individual galaxy within the cluster . . . you'll only see the line of sight velocity component. Do you understand what I am saying? If it's moving completely tangentially to the line of sight, you will see no Doppler effect from that motion. You'll only get the cosmological redshift of the cluster.

Though the jargon is formidable, that explanation actually makes sense.

At first I thought I had found a kindred spirit. Actually, though, the speaker was John Hartnett, who writes books with titles like *Dismantling the Big Bang: God's Universe Rediscovered*. He was speaking later in the conference on the topic "Starlight, Time and the New Physics." He had written a book with the same title, offering a novel explanation for the problem of distant starlight in a young universe.[2] Thankful for his support, I went to the store and bought a copy.

After the talk Matthews approached me, and we had a surprisingly cordial conversation. We did not talk much about astronomy, as I recall, but mostly exchanged personal stories. When I subsequently wrote an account of the conference for my blog, Matthews turned up in the comments. He was gracious enough to write this:

> First and foremost, I would like to reiterate how much I enjoyed our conversation at the ICC. You are a thoroughly delightful person. This may sound strange to you (it seems strange to me), but after we talked, I was filled with thanksgiving to God for what a delightful creature of His you are and what a blessing it was to converse with you (I don't recall that ever happening before).

> I certainly hope that some day we will meet again in person and have much more time to talk than we did that night. You are an absolute gem of a person.

I am happy to reciprocate the thought from my side, minus the religious language of course. I enjoyed our conversation thoroughly, and I think we could easily be friends despite our differences of opinion. Moments like these must be among the reasons I persist in attending these conferences. For all my disagreements with their views, I like being around people who are fired up about big questions.

34

Things I Learned at the Banquet

The ICC conference ended with a banquet dinner. Shortly before it began, I became involved in one of those conversations where several creationists pepper me with talking points without giving me a chance to reply. (They explained to me that materialism was plainly a fallacious philosophy, since the laws of logic are true but they are not material.) By the time we had finished, only a few empty seats remained. I grabbed the nearest one, at a table with seven other people. Three of them were high-school students. The remaining four comprised two pairs of married couples. One was older, roughly in their sixties. The other couple looked to be in their forties.

The ensuing conversation played out along familiar lines, without a trace of rancor or ill will from either side. We laughed at each other's jokes and generally agreed that this issue could benefit from everyone's calming down. I was interrogated a bit about being an atheist, but not at the level of abstract philosophy. Instead the questions were more personal. When did I know I was an atheist? Why did I care about anything if I did not believe in God? (That is precisely how the question was phrased.) They told me they could not imagine the loneliness of trying to live without God. I was reminded of a statement by philosopher Julian Baggini: "To those who say they cannot imagine the possibility of God not existing, I say try a little harder" (Baggini 2003, 99–100).

As usual in such things, the high-school students made for the most interesting conversation partners. One of them eventually asked me for an evolution reading list, and I jotted down a few books and websites for her. When I later learned she was the daughter of one of the conference organizers, I worried she might get in trouble if her parents found that list on her.

Of course, the old insularity was never far below the surface. At one point, apparently having decided the topics of evolution, creationism, religion, and morality were insufficiently controversial, we wound up talking about abortion. I commented that even Christians frequently disagree on important moral issues.

This high-school student became very cross. "All Christians are pro-life," she snarled.

I was too surprised by so blunt a statement to devise a better reply than mere contradiction. "No, that's not true. There are many pro-choice Christians."

Impressed by my strategy of contradiction, she deployed it right back at me. "All Christians are pro-life," she repeated.

Exasperated, I replied, "I personally know people who are both Christians and pro-choice."

I should have anticipated her reply. "Then they are not really Christians," she said, "because all Christians are pro-life."

"Why do you say that?" I asked.

"It's in the Bible!" she said, and directed me to a handful of Bible verses. One of them was Jeremiah 1:5, "Before I formed you in the womb I knew you, and before you were born I consecrated you; I appointed you a prophet to the nations." This is taken as evidence that the fetus has the same moral status as a human being. That Christians might disagree about the interpretation of such verses, or think that the Bible should not be the basis for public policy, never seems to have crossed her mind.

I was struck, though, by the fact that she had those Bible verses at her fingertips. Her parents would have been pleased, no doubt.

The banquet speaker was Dr. Andrew Snelling, a geologist. His two-volume work *Earth's Catastrophic Past*, totaling more than a thousand pages, is the most serious attempt at legitimizing young-Earth creationism of which I am aware. His topic for the evening: "The Creation Model: Its Past, Present and Necessary Future."

The past looked pretty good. Snelling focused on the success of *The Genesis Flood*, by John Whitcomb Jr. and Henry Morris. This book, published in 1961, marked the beginning of the modern creation-science movement. In Snelling's view, it was effective both scientifically and apologetically.

This was contrasted with the less happy present. Though Whitcomb and Morris paved the way, creationists had been less than fully successful in developing their ideas:

> Through much of our history the emphasis has been mainly anti-evolution, by attacking the evolutionary model, by developing and using scientific evidences, and by advocating our position fundamentally from science. Now, I'm not saying that's necessarily bad, but that's been our emphasis. Very much on the science, very much attacking the evolutionary model. And it needs to be attacked, in that sense. But not as much of our energy and resources have actually been put into building the creation model. Not simply as a viable scientific alternative, but rather as a model in itself.

The forlorn tone with which he said this prompted some sympathy from me. Surely you have had the experience of thinking you were the voice of reason among a group of highly confused people. Recall the frustration you felt trying to make yourself understood and your anger toward those who, in their ignorance, mocked you for your beliefs. I would imagine that people like Dr. Snelling feel that way most of the time, at least when contemplating their professional interests. Were he

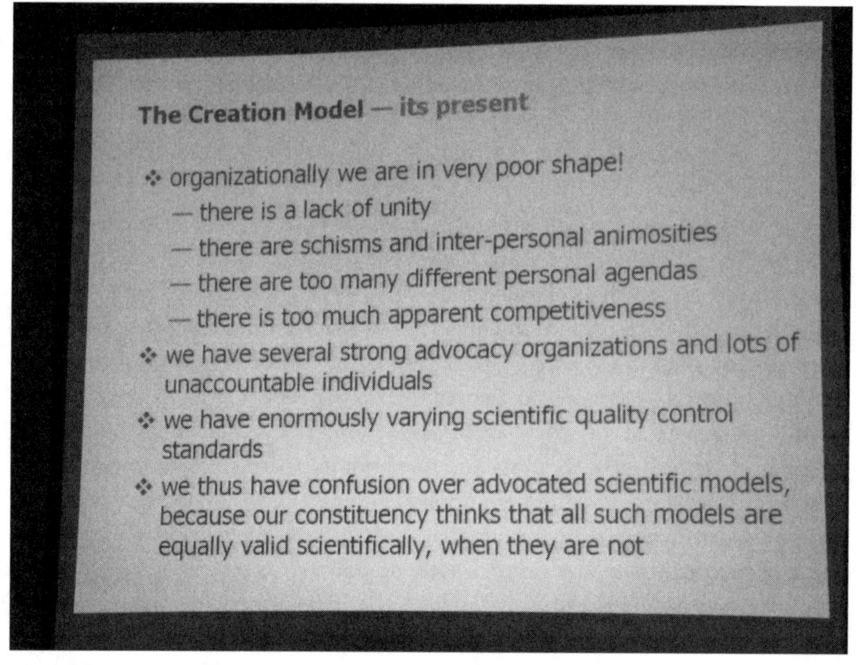

FIGURE 34.1 An encouraging slide, from the banquet presentation.

to walk into a conference of geologists and ask them to take his ideas seriously, the response would not be printable.

The most interesting moment came later, when Snelling said this:

> What if there was absolutely no evidence that the universe was young? No scientific evidence the universe was young. Would you still believe that it was young? Why? Because God's Word teaches it. That's the only reason you need to have to believe the universe is young. God's Word says it, therefore I believe it. That's not to say the evidences are not important. Of course they are. Because we're commanded to have a reason for the hope, and to give reasoned answers for what we believe and why we believe it.[1] But we must always remember our biblical foundations.
>
> So often we fight over the scientific evidence, but are we winning by leaving out our biblical foundations? Too much of our creation apologetics has therefore been based on the evidence alone. We need to keep arguing from the level of worldviews. Because ultimately the problem that people have is spiritual, the deliberate rejection of God's Word.

This, you might say, is the party line among creationists. Young-Earth geologist Kurt Wise echoes the thought:

> Although there are scientific reasons for accepting a young earth, I am a young-age creationist because that is my understanding of the Scripture. As I

> shared with my professors years ago when I was in college, if all the evidence in the universe turned against creationism, I would be the first to admit it, but I would still be a creationist because that is what the Word of God seems to indicate. Here I must stand. (Ashton 2000, 355)

Of course, most scientists do not agree that creationism's big problem is its excessive reliance on evidence. They would say the problem is that the evidence has, indeed, turned against creationism, but that people like Snelling and Wise refuse to accept it.

EPILOGUE

So ended the ICC, and so ends my story. I have no more profound finale to offer than that. My time with the creationists has had its ups and downs, but on the whole I have enjoyed it. I am grateful to them for their general tolerance of my presence at their gatherings, even if they were not always pleased I was there.

As it happens, I don't just attend creationist conferences. I have also had interesting moments among people whose religious faith is no less heartfelt, but whose feelings toward evolution are considerably more cordial.

In June 2009 I was in Cincinnati, Ohio, for the ninth North American Paleontological Convention. I attended a session called "The Nature of Science and Public-Science Literacy." Some of the talks provided food for thought regarding the best way of communicating science. Oddly, though, several others comprised a subconference on Christian apologetics.

The session abstracts made for interesting reading. One speaker informed us that:

> Darwin made another voyage in his life, from religious belief to skepticism. Darwin never rejected belief in God; he explicitly disavowed atheism. His agnosticism did not follow from his scientific studies but rather from the death of his beloved daughter Annie in 1851. Despite the cacophony of best-selling scientific writers such as Richard Dawkins, evolution co-exists as comfortably with religious belief as with atheism.

A different speaker provided the following characterization of the lingering suspicion of evolution among some Christians:

> Darwin's *Origin of Species* was greeted with a variety of responses among 19th Century Protestants. Some enthusiastically embraced it, integrating it into the core of their faith. Others, such as many Evangelicals and most Fundamentalists, opposed evolution on what they considered to be scientific grounds. Their opposition has continued for at least three reasons. 1) Current opponents believe that evolution is a theory or a hunch, and therefore it is not factual. 2) Some current opponents cling to the "orthodox" text, as though it were the latest word on the science, much as they might treat the Bible, ignoring 150

> years of development in diverse areas of biology. 3) Still others believe that "godless, materialistic" science, as they characterize it, has eclipsed faith throughout society, and they have set out to create a new evangelistic tool, "Creation Science," to encourage conversions to the faith (especially with youth).

Another offered some reasons for the antipathy, from both atheists and more conservative religious folks, towards theistic evolution:

> Most creationist critics of theistic evolution object to one or both of these characteristics of the view: 1) its reliance on naturalistic explanation, a feature common to all scientific theorizing; or 2) its embrace of "random" causal events, a feature common to myriad scientific explanations. Most atheist critics of theistic evolution object to its openness to supernatural explanation, a feature of religious belief in general. Such criticisms, valid or not, fail to address anything specific to theistic evolution.[1] To better understand the controversy surrounding theistic evolution, I propose that critiques of the position be considered in light of a lesser-known position we may (with tongue in cheek) call "theistic embryology." Theistic embryology describes the thinking of those who simultaneously express Christian belief and affirm basic theories in human developmental biology. . . . Most critiques that claim to identify weaknesses in theistic evolution make arguments that are equally damaging to "theistic embryology." . . .

Two other supporters of theistic evolution based their arguments on methodological naturalism (MN) (as we discussed in chapter 20). "Individuals on both sides of the public 'creation—evolution' debate have promoted a warfare view of science and faith based on a fundamental confusion of methodological naturalism with philosophical naturalism or materialism," wrote one. The other was even more blunt, "Science cannot say that the natural world is all there is, for MN forbids scientific discussion of entities beyond nature."[2]

Much of what transpired during the talks themselves would not have been out of place beneath a revival tent. One speaker boasted, to the audience's delight, that even if science provided an adequate naturalistic account of every facet of existence his faith would not be affected in the slightest. Atheists were bashed relentlessly for improperly mixing science with religion. This, recall, at a paleontology conference.[3] The general view was ably summarized in a talk bearing the stark title "Who Should Speak for Evolution: Atheists or Theists?" Little doubt was left regarding the speaker's answer. In his abstract he wrote: "In order to be helpful in support of science education, rather than just inflaming the controversy, atheists have to decide which they care about more: making our schools safe for evolution, or ridding the world of religion. Trying to do both at once alienates religionists who are their potential allies in supporting good science."

Certainly it was all a big departure from what I had heard from the creationists. That aside, there is much that is objectionable in these abstracts. Let us consider

Darwin's views on religion, for example. It is true that he was an agnostic and not an atheist, but to say that his religious skepticism was not the result of his scientific studies is overly simplistic. For example, in his posthumously published autobiography he wrote, in the context of explaining his religious skepticism:

> Although I did not think much about the existence of a personal God until a considerably later period of my life, I will here give the vague conclusions to which I have been driven. The old argument of design in nature, as given by Paley, which formerly seemed to me so conclusive, fails, now that the law of natural selection has been discovered. . . . There seems to be no more design in the variability of organic beings and in the action of natural selection, than in the course which the wind blows. Everything in nature is the result of fixed laws. (Darwin 1958, 87)

In a letter to biologist Asa Gray, Darwin wrote,

> But I own that I cannot see, as plainly as others do, and as I should wish to do, evidence of design and beneficence on all sides of us. There seems to me too much misery in the world. I cannot persuade myself that a beneficent and omnipotent God would have designedly created the Ichneumonidae with the express intention of their feeding within the living bodies of caterpillars, or that a cat should play with mice. (quoted in Miles 2001)

It is clear, then, that the failure of Paley's version of the design argument, coupled with the rampant cruelty and suffering in nature, each made manifest to Darwin as a result of his scientific studies, played a significant role in his religious skepticism. And since both of these points are front-and-center for anti-evolutionists today, it behooves us to keep them in mind when discussing Darwin's views in the context of modern debates.

It is likewise misleading to suggest that some nineteenth-century Protestants "enthusiastically embraced" evolution. As we saw in chapter 14, even among Christians inclined towards accepting some of Darwin's ideas, it was all but universal that natural selection, conceived of as a nonteleological mechanism acting on chance variations, was unacceptable, not just for scientific reasons, but for specifically religious reasons as well.

This is relevant to assessing the three reasons given by the speaker for the persistence of anti-evolutionism today. It is certainly true that creationists are misinformed regarding the evidence for evolution, often employ an untenable approach to the Bible, and see creation science as an evangelistic tool, but none of these points really capture what is at the heart of their hostility towards evolution. In the end they are opposed to evolution because they perceive, correctly in my view, that evolution severely challenges central tenets of Christianity.

Moving on, the speaker who argued that creationists object to theistic evolution because it invokes naturalism and randomness has also missed something important. There is nothing unreasonable in saying that naturalism and randomness are acceptable for some questions but not for others. If their argument was,

"We object to theistic evolution because it involves naturalism and randomness," then it might be appropriate to reply, "But you accept modern embryology, which also has those features." Instead the creationist argument is, "We object to theistic evolution because we believe, for both religious and scientific reasons, that the appearance of humanity on Earth cannot be understood naturalistically. It involved direct, supernatural intervention."[4] The relevant distinction between embryology and evolution is that the latter, unlike the former, addresses questions relevant to our understanding of humanity's place in nature and intrudes on territory on which religion had previously staked a claim.

As for the speakers who brought up MN, we can reiterate our earlier point that restricting science entirely to naturalistic causes in no way implies that religion has nothing to fear from science. Science gives us increasingly accurate pictures of how nature works. If your religion likewise addresses aspects of the natural world, then we clearly have the possibility of conflict. To someone who says, "You shouldn't worry about science conflicting with your faith, since science is purely naturalistic and therefore cannot address supernatural entities," a creationist could reasonably reply:

> Yes, thank you, we understand that science (as currently conceived!) is purely naturalistic. But our religion teaches that God is just and loving, and that makes it hard to believe He would do His creating by a mechanism like natural selection. Our religion also teaches that humanity plays a central role in Creation, and that is likewise hard to believe in the light of evolution.

Theistic evolutionists, obviously, have their counters to offer against these assertions, and people must decide for themselves whether those counters are adequate for defusing the challenges. My point is simply that directing creationists to the methodological limitations of science in no way addresses any of their deeper concerns.

If all I knew about this topic came from this one conference session, I would think that opposition to evolution is the exclusive domain of a handful of religious extremists, egged on by a few equally extreme atheists. My experiences among the creationists do not at all confirm this view. To the extent that confusions about the nature of science, dogmatic support for biblical literalism, or backlash against outspoken atheists play any role in the popularity of anti-evolutionism, they are distant side-issues. The main issues are so much simpler and, sadly, so much more intractable than that. Evolution challenges the claims of the Bible, refutes Paley's version of the design argument, exacerbates the problem of evil, and seems to diminish human significance. Any one of these points is worrisome, but all four together amount to a strong cumulative case against the possibility of harmony between evolution and Christianity.

In 2001, philosopher Michael Ruse published a book entitled, *Can a Darwinian Be a Christian?* He presented his answer in the book's epilogue: "Can a Darwinian be a Christian? Absolutely! Is it always easy for a Darwinian to be a Christian? No, but whoever said that the worthwhile things in life are easy? (Ruse 2001b, 217)."

Though I have made my own sympathies perfectly clear, I do not propose to give a definitive answer to Ruse's question. Instead I ask simply that we recognize it as a matter of opinion and not of fact. How you answer depends on what you believe is central to Christian faith and on what you consider it plausible to believe.

Ruse continued with, "Is the Darwinian obligated to be a Christian? No, but try to be understanding of those who are" (ibid. 217). I would ask that the same understanding be extended to those who find the reconciliation too difficult to manage. Once you have acknowledged that evolution forces a profound rethinking of traditional faith—and how can you not?—why is it unreasonable to conclude our reappraisal with the finding that one of the two systems must simply give way?

Upon entering the world of theistic evolution, you will encounter brilliant scholars from science, philosophy, and theology writing at book length just to defend the *possibility* of reconciling evolution with Christianity. To this I can only reply that if you are inclined towards atheism, or tend to think that human significance in the cosmos has been dramatically overstated, then it is hard to imagine anything science might have discovered more congenial to your worldview than evolution by natural selection. For the thoughtful Christian, evolution poses a challenge, one he or she must work hard to defuse. The atheist, by contrast, responds with perfect equanimity. He or she just smiles and says, "Of course. How else could it be?"

My argument is politically inconvenient. In all of the various squabbles between evolution and creationism in the classroom, such successes as evolutionists can claim have come in large measure because of the support of religious people. Speaking as someone who cares passionately about science education, this puts me in the awkward position of having to be happy that so many people disagree with my view about the relationship between science and religion. So be it. At a personal level I think very highly of many of the creationists I have met, but ultimately we are on opposite sides. They are wrong about important things, and their ideology must be vigorously opposed whenever it attempts, in even the slightest way, to influence public policy or education.

NOTES

Chapter 1

1. I would also note that the phrase "algebraic division/graphing" is very odd. It is hard to imagine what "nonalgebraic" division could mean, and at any rate it is unclear why division is naturally paired with graphing. By itself this is not such a big deal, but when similar examples appear on page after page the impression is created that the writer does not really understand the subject.

2. There is a large literature devoted to describing evangelical culture. For discussions of Christian home-schoolers specifically, see Kunzman (2010) and Rosin (2007). For creationist culture, see Eve and Harrold (1991) and Toumey (1994). For more general discussions of evangelical culture, written from the perspective of outsiders immersing themselves within it, see Roose (2009) and Welch (2009). More detached discussions are available in Balmer (2006) and Sheler (2006)

Chapter 5

1. The problem being that if God knows everything that will happen in the future, then he also knows what decision I will make in any given situation. But since God can never be wrong, that suggests I am not really free to make any choice I please.

2. For a readable introduction to the philosophical question of justifying morality without reference to God, I recommend Sinnott-Armstrong (2009).

3. "The Tea Party" is a reference to an American political movement that has come to prominence since the election, in 2008, of President Barack Obama. They are extremely right-wing, and in the opinion of most knowledgable commentators badly misinformed about American history and terribly misguided in their policy recommendations.

Chapter 6

1. In recent years biologists have come to appreciate the significance of "horizontal gene transfer" (HGT), especially among unicellular organisms. Normally we think of genetic material being passed from parent to offspring, a process we might think of as "vertical gene transfer." In certain cases, however, genetic material can be passed between organisms that do not share a parent-offspring relationship. Among complex, multicellular organisms, HGT seems rare to nonexistent, but it is common among simpler organisms. This suggests that the metaphor of a "tree," which is based on the idea that genetic transfer is vertical, is inapt for the very earliest stages of evolution. It has been suggested that something more like a bush is an appropriate image. This is an ongoing area of current research, but it is a level of detail that is not so important for the discussions to come.

Chapter 7

1. I say "roughly speaking" because even within the somewhat narrow confines of young-Earth creationism there are significant divisions and differences in emphasis. Some of these distinctions are discussed by geologist Richard Peters (2009).

2. Frye (1983) is one of many useful references here.

Chapter 8

1. Let me remind you that I am simply describing things as YECs see them. I certainly am not necessarily endorsing the correctness of any of their analyses.

2. In light of how I introduced this section, I should mention that Morris's article is entitled, "The Bible *Is* a Textbook of Science." Reading past the title, however, makes it clear that he is making the same point I am emphasizing here.

3. In a more recent discussion, Ken Ham (2011) expressed similar views: "Many great men of God who are now with the Lord have believed in an old earth. . . . Scripture plainly teaches that salvation is conditioned upon faith in Christ, with no requirement for what one believes about the age of the earth or universe."

Chapter 9

1. This slogan is especially prevalent in the work of Answers in Genesis, though the principle is at least implicit in most YEC literature.

2. The terms "microevolution" and "macroevolution" are sometimes used by biologists as well. In this usage, these terms generally refer, respectively, and somewhat informally, to evolution below and above the species level.

Chapter 10

1. My description of Kerby's presentation is based on the DVD version of the talk, which differed slightly from the version presented at the conference.

2. Creationists deny that it is possible to arrange fossils chronologically, of course, but for the purposes of this section I will take it for granted that the various dating measures used by scientists are sound. Dalrymple (2004) provides a readable introduction to scientific-dating methods.

Chapter 11

1. This, and all subsequent quotations of Silvestru in this chapter, have been transcribed from the DVD presentation of his talk. What is presented on the DVD is very nearly identical to what I saw at the conference.

2. We should note that for reasons related to certain technical issues in systematics (the branch of biology devoted to working out the proper evolutionary relationships among modern organisms), the term "mammal-like reptile" is considered obsolete among paleontologists. Professionals nowadays prefer the term "stem-mammals."

3. By far the best single-volume reference for this subject is Prothero (2007). See also Miller (1999), Coyne (2009b), and Dawkins (2009) for readable discussions of transitional forms in evolution.

4. As in other chapters, I have relied on the DVD version of Bell's talk for direct quotations. The DVD presentation was very nearly identical to the version delivered at the conference.

5. It should also be noted that creationists themselves do not always agree on the proper classification of these fossils. Even as they claim that all such fossils are either clearly ape or clearly human, different creationist authors come to different conclusions as to which is which. Have a look at Foley (2008) for the details. That creationists themselves are unable to make a clear determination as to which are the apes and which are the humans is itself strong evidence of the transitional nature of these fossils.

Chapter 12

1. The quotations in this section have been drawn from the conference program, but Gitt has written a book presenting the same ideas (Gitt 2006). As it happens, this was the book that Ken Ham recommended to me in chapter 1.

2. This is just one of several mechanisms that can increase information. See Maynard Smith and Szathmáry (1995, 10).

3. The refusal to acknowledge that it is the combination of duplication and divergence that can lead to an increase in genetic information, and not just duplication by itself, is a common error in anti-evolution writing. For example, in his book *The Wedge of Truth*, ID proponent Phillip Johnson writes, "When two rabbits reproduce there are more rabbits, but there is not any increase in information in the relevant sense. If you need to write out the full text of the encyclopedia and have only page one, you cannot make progress toward your goal by copying page one twenty times (Johnson 2000, 59)." Of course, this statement simply fails to understand the relevant genetic processes. A more detailed discussion can be found in Rosenhouse (2001).

Chapter 13

1. The direct quotations in what follows are taken from Kerby's 2006 book *Remote Control: The Power of Hollywood in Today's Culture*, which is essentially an extended version of the talk I attended.

2. As an aside, the characters in *Rope* were based on the real-life case of Leopold and Loeb, who in 1924 committed a murder comparable to the one in the film. At trial they were represented by Clarence Darrow, who later became lead counsel for the defense in the Scopes "Monkey" trial. Darrow entered a guilty plea, but successfully argued against the death penalty. His argument was based in part on the idea that we are all subject to the biological forces that created us, and are therefore not completely responsible for our actions. A common fear among creationists is that evolution leads to the demise of clear moral thinking. You can imagine, therefore, what they think of Darrow!

Chapter 14

1. By "the basic argument" I mean simply the idea that the fact of organismic complexity in some way justifies the conclusion that God (or some other intelligent designer) exists. Philosophers have devoted considerable effort to teasing out the exact logical form of the design argument, as presented by different authors. For example, should the design argument

be seen as based on analogy (animals are like machines, and machines are built by intelligent designers)? Or is it better viewed as an inference to the best explanation (organismic complexity is easy to understand given the assumption of an intelligent designer, but is hard to understand under any proposed rival hypothesis)? Of many possible references, Sober (2004) provides a readable discussion of these issues.

2. They could point to scholars like William Whewell, who had argued along these lines in some of his writing. Paley's particular version of the design argument was somewhat out of fashion, even before Darwin published his theory.

3. Interestingly, Paley himself acknowledged this point:

> My opinion of astronomy has always been, that it is *not* the best medium through which to prove the agency of an intelligent Creator; but that, this being proved, it shews, beyond all other sciences, the magnificence of his operations. The mind which is once convinced, it raises to sublimer views of the Deity, than any other subject affords; but is not so well adapted, as some other subjects are, to the purpose of argument. (Paley 2006).

4. In emphasizing the spiritual and devotional aspect of Paley's argument, I am not suggesting that the intellectual aspect is unimportant. By refuting Paley's argument, Darwin really did knock the legs out from beneath one of the strongest arguments ever devised for believing in God. The relevance of this to modern discussions about science and religion has been ably expressed by philosophers Daniel Dennett (2006), John Dupré (2003, 41–62), and Steve Stewart-Williams (2010, 45–53).

5. You should recall that in Darwin's time the physical basis for heredity (genes) was unknown. A proper theoretical development of natural selection was not possible in the nineteenth century, but had to wait for substantial progress in genetics.

6. This general rejection of Darwin's mechanism raises an interesting question. We have seen that many people accepted common descent (or at least did not find it theologically problematic) while rejecting natural selection. What evolutionary mechanism did they propose in its place? As documented by Moore (1979, 125–173), there were a variety of suggestions on offer, many of which were mystical or even supernatural. In saying that such people accepted common descent, we should be clear that they were not endorsing a fully naturalistic account of natural history. I find this significant, because anyone advocating such a view today would be considered an anti-evolutionist, if not an outright creationist. Intelligent-design proponent Michael Behe, who accepts common descent but argues that some reference to intelligent activity is necessary for understanding natural history, is an example.

7. In a recent essay, YEC Terry Mortenson lamented this fact, writing: "For thousands of years Christians have faced the danger of being deceived by Satan (2 Corinthians 11:3). Sometimes in the process of defending the Christian faith, respected Christian leaders have unknowingly compromised with error. Surprising examples include founders of modern fundamentalism" (Mortenson 2010).

Chapter 15

1. During the time I was writing this book, I mentioned this story to my father. He told me that, as it happens, that intersection has since been redesigned precisely to fix the problem I have described.

2. Gould has received some criticism for casually discussing what a "sensible God" would or would not do (Nelson 1996; Sober 2008, 126–129). If Gould intended his assertion as the conclusion of a rigorous theological argument, then I agree that the criticism has merit. A more reasonable reading, however, is that Gould was simply using the phrase "a sensible God" as a catchy way of saying that "odd arrangements" are hard to understand from the standpoint of a human engineer, but easy to understand as the endpoint of a blind historical process

3. For discussions of this principle specifically with respect to human genetics, I recommend the books by Avise (2010) and Fairbanks (2008).

4. Creationists deny this. Much of their literature consists of heroic efforts to explain why, no matter how convoluted a structure might seem, it nonetheless reflects the wisdom of an omnipotent designer. Given the sheer profligacy of engineering oddities in nature, I would say they have their work cut out for them.

5. A classic example of this attitude is found in Carter (2001). A spokesperson for the Discovery Institute, the leading think tank promoting ID, commented on a television documentary about evolution by saying, "The final episode paints a picture that the only critics of Darwinian theory are these guitar-strumming hillbillies in Kentucky who are creationists, and that's just not true. We're glad we're not part of that stereotype." For further discussion of this statement, see Scott and Branch (2003).

6. See Purdom (2005) for a published version of the talk. It is interesting that some defenders of theistic evolution have made precisely this same point against ID (Ayala 2008; Hess 2010). We shall discuss this further in chapter 24.

7. The books by Larson (1989), Nelkin (1982), and Singham (2009) all provide helpful surveys of the legal history of evolution/creation disputes in education.

8. Only four of the five justices on the court heard the case. Two of the justices comprised the majority from whom I quoted. A third justice agreed that the Butler Act was constitutional, but only if interpreted so narrowly as to remove nearly all of its force. The fourth justice found the law unconstitutionally vague.

9. It should be noted that the framers of the Louisiana law, realizing that the Arkansas law was unlikely to survive a constitutional challenge, greatly toned down their language. This was really the birth of the legal strategy that later became known as ID (Matzke 2009).

10. For detailed discussions of the frequently misleading media coverage of this issue, see Mooney and Nisbet (2005) and Rosenhouse and Branch (2006).

11. For detailed treatments of the Dover trial, I recommend the books by Humes (2007) and Lebo (2008).

Chapter 16

1. Trilobites are an extinct group of ancient arthropods, roughly similar to modern horseshoe crabs. Brachoipods are a group of marine organisms characterized by certain properties of their shells, similar to modern mussels and scallops. Ammonites are another group of extinct, marine invertebrates, closely related to modern squid and octopuses.

2. Stromatolites are layered structures formed when certain minerals become trapped by large colonies of bacteria. Fossil stromatolites record the activity of some of the oldest forms of life on Earth. The Ediacaran and Vendian eras in Earth history preceded

the Cambrian. Thus, Ward is here referring to mid-twentieth-century fossil finds that provided evidence of the sort of life that existed prior to the forms recorded in the Cambrian rocks.

3. In this context, "metazoan" is essentially synonymous with "animal."

4. Johnson is taking advantage of the same equivocation we discussed in chapter 11. The question of whether the fossils confirm the major predictions of evolutionary theory is separate from the problem of trying to determine specific lines of descent.

Chapter 17

1. Let me mention that Wells's arguments about embryology are badly mistaken. See Gishlick (2008) and Matzke (2004) for helpful discussions.

2. As a supplement to this chapter, the article by Harding (1987) provides an interesting discussion of the rhetoric of fundamental Baptist conversion.

Chapter 18

1. I am thinking specifically of conversion experiences. The term "religious experience" covers a wide variety of different phenomena, of which conversions are only one. What I find compelling about conversion experiences is the tangible evidence they leave behind. It is very easy for someone to claim to have had a religious experience, especially when they are immersed in a culture that values such things. But if a person fundamentally changes his life as the result of the alleged experience, that is good evidence that *something* significant took place. Whether that "something" involves contact with a supernatural realm is, of course, highly debatable.

2. For the true chess fans, it seems that white had a very strong move with 1. Qe2! (instead of 1. Qe4). The main point is that after 1. . . . Rxf4 2. Qb2+ Kf8 white plays 3. Qb8+ and grabs the rook on f4. So black would have to try 1. . . . Kh6 2. Kg3, but his pieces are very uncoordinated. White probably has enough tactical tricks at his disposal to make it very difficult for black to hold. The endgame is quite complicated though. We should also note that, in the diagram position, black has another resource in 1. . . . Rxf4! The pawn ending after 2. Qxf4 g5+ 3. Qxg5 Nxg5 4. Kxg5 is a straightforward draw. Of course, the move played in the game wins outright, but it is interesting that after white's blunder black has two tactical possibilities.

3. Philosopher Theodore Drange's *Nonbelief and Evil* (1998) is an excellent book discussing both arguments.

4. For a more thorough discussion of Swinburne's arguments about the validity of religious experience, written from a skeptical perspective, please consult the essay by Richard Gale (1994).

5. Other objections to Alston's argument can be found in Fales (2004).

6. William Alston is one of the most prominent defenders of the validity of religious experience, but space restrictions make it impossible for me to discuss his ideas further. The most comprehensive presentation of his views is found in his book *Perceiving God* (1991). Alston's arguments have been strongly criticized by a number of philosophers, with Kretzmann (1994) being one helpful reference among many.

Chapter 19

1. As always, the literature in this area is vast. Among philosophers, William Lane Craig has been the most passionate defender of the Kalām argument, while Quentin Smith has been one of its most vocal critics. They hash out many of the relevant issues in their jointly authored anthology (Craig and Smith 1995). I have also found the following references helpful: Mackie (1982, 81–101) is critical of cosmological arguments generally, and the Kalām argument specifically. Craig (1984) defends the argument against Mackie's criticisms, but Oppy (1991) then rebuts Craig. Oppy (2002) provides additional criticisms. Craig (1997) provides a further defense of the argument. Smith (2006) is a cheeky attempt to use Kalām-style reasoning to advance atheism. I also recommend the book by physicist Sean Carroll (2010). He does not discuss cosmological arguments specifically, but he does explicate many of the relevant scientific issues. Finally, since most of these references make for heavy reading, you can consult Barker (2008, 130–144) for an impressively lucid, and very readable, treatment of the same issues.

2. Greene (2011) is a popular-level exposition of some of the reasons multiverse speculation is currently very fashionable among physicists. Stenger (2011) provides ample reasons for skepticism regarding the fine-tuning argument. Manson (2000) argues that the whole idea of "fine-tuning" is too vague to serve as the basis for any argument that concludes the existence of "extra-cosmic" entities. McGrew, McGrew and Vestrup (2001) argue along similar lines, casting doubt on whether it is meaningful to discuss probabilities in the context of fine-tuning. The anthology edited by Carr (2007) offers a variety of scientific and theological perspectives, but be warned that most of this book makes for heavy reading.

I have emphasized that if our universe is part of a larger multiverse, then it is no longer remarkable that at least one universe would have the properties necessary to support life. A stronger, and more controversial, claim is that the fine-tuning of our universe is actual evidence for a multiverse. This is really a philosophical question, since it hinges in part on difficult questions regarding the nature of "evidence." Hacking (1987) argued against the claim, but he was challenged by Leslie (1988), McGrath (1988), and Whitaker (1988). White (2000) sided with Hacking. Holder (2002) and Manson and Thrush (2003) challenged White. More recently, Draper, Draper, and Pust (2007) sided with White. No doubt the debate will continue.

3. Of course, at this point religious fish would want to know where the filter came from, and scientist fish would reply that, while its origins are mysterious, the origins of a filter-making deity would be even more mysterious.

4. They key phrase here is "in principle." Just to be clear, I do not believe the specific arguments made by ID proponents are at all convincing. The phenomena they point to as requiring explanations based on intelligent agents do, in fact, have adequate naturalistic explanations.

5. As an aside, for a book-length discussion of the place of science in society, I recommend the book by Levitt (1999).

Chapter 20

1. There is also another legal standard known as the endorsement test, presented in the 1984 Supreme Court case *Lynch v. Donnelly*. It holds that a government action is unconstitutional if it would create a perception in the mind of a reasonable observer that the

government is approving or disapproving of religion by taking that action. In the *Dover* trial, Judge Jones found that the ID policy ran afoul of this test as well.

2. The third prong requires that the action not represent an "excessive government entanglement" with religion.

3. Several other philosophers have defended the judge's decision, and thus indirectly Ruse's expert witness testimony (Gross 1983, Oppy 2003).

4. For a detailed discussion of this point, see Dawes (2009, chs. 5–7).

5. Pennock has replied to Sober's example:

> Presumably he thinks that the ordinary observation that not everything in nature is purple shows that the hypothesis is false because an omnipotent being with such a desire would surely have made the world so. If we are thinking of this in terms of a naturalistic understanding of notions such as "a being," "desire", "above all," "nature," "to be," "purple," and so on, that test may be perfectly fine. But what can we say when we treat the hypothesis supernaturally? Might not all of nature now indeed "be purple" in its noumenal substance, irrespective of its accidents, as wine purportedly becomes God's blood without observable change in the miracle of the Eucharist? Is it even possible for God (to specify the being behind this generic talk of omnipotent supernatural entities), to want such a thing "above all"? . . .
>
> One could easily continue, but let us not belabor the point—it is only under a tacit naturalistic reading that the testability of hypotheses containing such concepts could be thought plausible. (Pennock 2011, 190)

I'm afraid I cannot fathom Pennock's point. He has certainly persuaded me that one can take Sober's simple example, apply bizarre definitions to the words, and transform it into a claim that is not testable, or perhaps even meaningful. The fact remains that Sober's example makes perfect sense when the ordinary meanings of the words are applied, and it is a clear example of a supernatural hypothesis that can be falsified by empirical data. For further discussion, see Sober (2010, 142–146)."

Chapter 21

1. The reference here is to an ongoing experiment in bacterial evolution carried out under the supervision of biologist Richard Lenski. Dawkins (2009) contains a lengthy discussion.

2. Also contributing to whatever intellectual heft the ID folks can claim is the work of mathematician William Dembski. He offers a variety of technical arguments based on probability and optimization theory. Since his arguments were barely mentioned at the conference, I have not found it convenient to discuss them here. See Dembski (2001) for a clear statement of his major claims. For replies and refutations, see Godfrey-Smith (2001), Perakh (2004), Rosenhouse (2002), and Shallit and Elsberry (2004).

Chapter 22

1. Both of these arguments are addressed in Isaak (2005)

Chapter 23

1. Philosopher Stephen Asma raised the same issue after his own visit to the museum:

> It's not quite accurate to call this evangelical center a "museum." It contains almost no "information," unless you count as information speculations on how Noah kept dinosaurs on the ark. It offers no new observations about nature, unless you think that inferring a Designer can be called observational. Unlike most other nature museums, it has no "research" component whatsoever. When I asked Mark Looy, vice president for AiG [Answers in Genesis] ministry relations, where the research labs and archive collections were located, he confessed that he didn't understand the question. "This is a museum," he finally said, chuckling. (Asma 2011)

Chapter 24

1. They certainly are not obviously contradictory, in the sense that they do not have the form "*X* " and "Not *X*." Defenders of the logical argument attempt to provide some further statement that is both self-evidently true and which, in conjunction with "Evil exists," implies that God does not exist. Useful discussions can be found in Mackie (1955), McCloskey (1960), Parsons (1990), and Rowe (2001).

2. But see the essays in Martin and Monnier (2003) for a counterpoint, especially part 2 of the book.

3. Especially important in this regard is Alvin Plantinga's 1971 essay, "The Free-Will Defense" (Plantinga 2001a). In it he refutes, successfully in the opinion of most philosophers, the logical problem of evil. I would note, though, that the essay consists of nearly thirty pages of dense, analytic philosophy. If you have to work *that* hard just to defend the logical possibility of God's existence in the face of great evil, then we should not be sanguine about the magnitude of the problem of evil.

4. Dembski (2009) offers an imaginative argument for rejecting this conclusion, suggesting that human sin could have affected the creation retroactively. Lack of space precludes a proper discussion of his rather implausible view.

5. The term "skeptical theism" is also used.

6. Philosopher Paul Draper raises an interesting point that has some relevance here. He notes that many defenders of theism argue that we are sufficiently cognizant of God's intentions and motives to conclude that special creation was theologically implausible even without evolution to deliver the final blow. Many of these same thinkers, however, resort to theistic skepticism when replying to the problem of evil. There would seem to be an inconsistency here. See Draper (1998) for a discussion.

7. Martin (1988) provides a more detailed discussion of this point.

8. I am indebted to Glenn Branch for suggesting this analogy to me.

9. Here's the full quotation:

> The total amount of suffering per year in the natural world is beyond all decent contemplation. During the minute it takes me to compose this sentence, thousands of animals are being eaten alive; others are running for their lives, whimpering with fear; others are being slowly devoured from within by rasping parasites; thousands of all kinds are dying of starvation, thirst and disease. It must be so. If there is ever a time of plenty, this very fact will automatically lead to an increase in population until the natural state of starvation and misery is restored. (Dawkins 1996, 132–133)

10. A possible escape here is to argue that even brilliant scholars can work with an incoherent idea without noticing. Mathematicians, for example, worked with set theory for

quite some time before realizing that Russell's paradox calls its foundations into question. But this hardly seems comparable to what Haught describes as theologically unthinkable. In the case of set theory, the paradox was very subtle and was irrelevant to most of the things mathematicians used sets for. By contrast, the notion of a (relatively) instantaneous creation was a central notion in Christianity for centuries and was integral to the stories of Adam and Eve and the Fall. Thus, I would say that Haught's blunt statement about an instantaneous creation being "theologically unthinkable" has as much credibility as creationist claims that scientists have overlooked elementary errors in evolutionary theory.

11. It is interesting to note that Southgate provides just three reasons for hypothesizing a pelican heaven. He points first to a handful of vague Bible verses ("profoundly enigmatic" is his phrase). He then points to an equally vague "general inference from reflection on the picture Scripture offers." Finally, he writes, "A conviction as to the goodness of God also requires that we invoke the doctrine of redemption, and postulate some sort of eschatological fulfillment for creatures. (Ibid. 82)" The reader can decide for himself whether these are compelling reasons.

12. For a more thorough discussion of how process theologians address the problem of evil, consult Whitney (1989, 47–57).

13. We shall discuss this further in chapter 30. Let me also protest Haught's casual assertion that those who see no directionality in evolution are "pessimistic." Personally I find the notion of humanity being a chance by-product of evolution to be far more uplifting than anything envisioned by traditional theology.

Chapter 25

1. This quotation is accurate. It can be found on page 127 of Gould's *Ontogeny and Phylogeny* (1977).

Chapter 26

1. We should mention that creationists would not entirely agree with Gilkey's characterization. They would note that the physical topography of the Earth changed dramatically as a result of Noah's flood, meaning it is different today from what it was moments after the creation.

2. In their more florid moments creationists describe the Bible as sixty-six love letters from God, written in heavenly handwriting.

3. The notion that the Bible contains inconsistencies, internal contradictions, and clear evidence of editing, obvious to virtually everyone who has read it seriously, is bitterly rejected by creationists. In reply they have published many books meant to show that all such objections are unfounded. Archer (2001) is an especially comprehensive contribution to the genre. Archer was an old-Earth creationist and defended the day-age approach to Genesis 1. More than once I have had YECs recommend the book to me (which, indeed, is why I bought a copy in the first place), but only with the warning that I ignore what he says about Genesis 1. At any rate, for a young-Earth approach to the same problem, try Ham (2010).

Chapter 27

1. The Bible describes the creation of Adam in Gen. 2:4-7:

"In the day that the Lord God made the earth and the heavens, when no plant of the field was yet in the earth and no herb of the field had yet sprung up—for the Lord God had not caused it to rain upon the earth, and there was no one to till the ground; but a stream would rise from the earth, and water the whole face of the ground—then the Lord God formed man from the dust of the ground, and breathed into his nostrils the breath of life; and the man became a living being."

The statement that "there was no one to till the ground," strongly implies that Adam was not drawn from a pre-existing hominid population. Moreover, the statement that "the Lord God formed man from the dust of the ground," just prior to breathing life into him implies that it is the origin of Adam's physical body, and not just the origin of his soul, that is being explained here. That notwithstanding, speculations about "pre-Adamite" human beings has a long history, inspired by a variety of political, scientific and theological considerations. Livingstone (2008) provides a thorough history of these ideas.

2. Philosopher Kenneth Kemp, in a paper defending ideas very similar to Alexander's, tries to circumvent this point by writing: "A theology in which the existence of a Chosen People is a central theme in salvation history can surely accommodate the existence of a Chosen Couple. God did not owe Adam and Eve's cousins a rational and therefore immortal soul. The hominization of Adam and Eve was a free gift (Kemp 2011, 233)." This response is too casual. A mother and father might not "owe" Christmas presents to their two children, but, having decided to give presents at all, most would consider it an injustice to give presents to one child but not the other. Likewise, once God decided to bestow the gift of a rational soul to at least some of his creatures, we can reasonably wonder whether it is just for Him to provide this gift to some but not others. Even aside from concerns about fairness, we might wonder why, if God desired a special relationship with His creatures, it would serve His purposes to single out just two people to receive the gift of a soul. This concern becomes more acute when we note that God cursed the ground for everyone solely because of the sins of the first couple. Surely, if God had gifted a large population with a rational soul, some would have made better use of the gift than others.

Moreover, Kemp's analogy to the idea of a Chosen People is problematic. Among Jews, a very common understanding of that notion is that the Jews are chosen by virtue of their unique willingness to enter into a covenant with God. On this understanding, the Jews alone accepted a gift that was offered to everyone. If this is correct, then Kemp's analogy clearly breaks down. Though I lack the space to explore the issue here, I would argue this is the only understanding of the idea of a Chosen People that does not raise serious issues about God's justice and goodness.

3. The *Enuma Elish* is a Babylonian creation myth.

4. For a counterpoint, albeit one I do not find persuasive, have a look at the paper by Jackelén (2008).

5. The question of Christianity's role in the development of modern science is fiercely debated. Creationists are fond of telling their audiences that science owes its very existence to Christianity. Some atheists argue instead that Christianity had nothing at all to do with it. This is one case where reality really is somewhere between the two extremes. See Efron (2009) for a helpful discussion.

Chapter 28

1. Yes, Jews also conduct Sunday schools

Chapter 30

1. Paleontologist Simon Conway Morris, who demurs from Gould's view, has written, "We may be unique, but paradoxically those properties that define our uniqueness can still be inherent in the evolutionary process. In other words, if we humans had not evolved then something more or less identical would have emerged sooner or later. Such an idea is dismissed out of hand by nearly all evolutionary biologists. . . . (Conway Morris, 2003, 196)" Biologist Jerry Coyne (2009a) and paleontologist Donald Prothero (2003), both of whom agree with Gould against Conway Morris on this issue, also affirm that Gould's view is the dominant one among evolutionary biologists.

2. Physicist Taner Edis provides a good discussion of these points. He writes, "The claim that randomness is a crack in the natural order that makes room for supernatural action is also wrongheaded. Far from allowing miracles without violating physics, such action would require a significant rupture in the present structure of physics. For it is nothing less than a claim that quantum mechanical events are *not* random" (Edis 2006, 45–48).

3. How plausible the argument seems does not appear to depend on one's views towards the compatibility of evolution and religion. Miller and Conway Morris both defend theism, but similar ideas have been defended by philosopher Daniel Dennett (Dennett 1995, 299–312), who is well-known for his outspoken atheism. (Though, to be clear, Dennett stops short of saying he believes humanlike intelligence specifically was inevitable.) From the other side, Michael Ruse, an agnostic philosopher who argues passionately for the compatibility of evolution and Christianity, is nonetheless critical of the argument (Ruse 2010, 221).

I also am not suggesting that this question is interesting and relevant only for its theological implications. Determining the relative roles of contingency and predictability in evolution is important for a proper understanding of biology. It also plays into the broader philosophical question of whether there are laws in biology in the same sense as there are in physics. Beatty (1995) has used Gould's contingency thesis as the foundation for an argument against the existence of biological laws. His view was subsequently criticized by Sober (1997) and Elgin (2006), who in turn were criticized by DesAutels (2009). Of course, there are many, many other references as well.

4. My analogy is imperfect since I have casually equated "contingency" with "sensitivity to initial conditions." As several philosophers have noted, there are different notions of contingency that can be relevant in different contexts. I have not found it necessary for my brief discussion to explore such details, but if you are interested you can start with Beatty (2006) and Ben-Menahem (1997).

5. Also relevant here is this statement, from Stephen Jay Gould:

> Our closest ancestors and cousins, *Homo erectus*, the Neanderthals, and others, possessed mental abilities of a high order, as indicated by their range of tools and other artifacts. But only *Homo sapiens* shows direct evidence for the kind of abstract reasoning, including numerical and aesthetic modes, that we identify as distinctively human. All indications of ice-age reckoning—the calendar sticks and counting blades—belong to *Homo sapiens*. And all the ice-age art—the cave paintings, the Venus figures, the horse-head carvings, the reindeer bas-reliefs—was done by our species. By evidence now available, Neanderthal knew nothing of representational art. (Gould 1989, 319–320)

Chapter 31

1. The reference to Georgia was about a 2006 court case that ensued when a local school board voted to put stickers in science textbooks warning people about evolution. The trial court judge found the stickers to be in violation of both the federal and state constitutions. A three-judge panel for the appeals court vacated that finding on the grounds that some of the facts on which the trial judge based his conclusions were not adequately supported by the trial record. This had the effect of returning the case to the lower court to be retried. The case was ultimately settled without a retrial, with part of the settlement leading to the removal of the stickers.

2. An example is the ID "documentary" *Expelled*, which featured stories from several scientists and educators who claimed their careers were damaged or ended because of their anti-evolution views. However, in every case the facts of the situation were relevantly more complicated than what was presented in the film. Visit the website "*Expelled* Exposed" at http://www.expelledexposed.com/ for the details (last accessed September 4, 2011).

3. In addition to the other errors in these passages, we should note that the main title of Darwin's 1859 book was *On the Origin of Species*, with no "the" before "Species."

Chapter 33

1. The paper in the conference procedings presented two other lines of evidence, one involving the cosmic microwave background radiation and the other involving patterns in gamma ray bursts. Suffice it to say that more mainstream astronomers would not agree with Matthews's interpretation of the data.

2. The problem is that we detect light from stars that are billions of light years away, implying that the light from those stars must have been traveling for billions of years to get here. If the universe is only a few thousand years old, then that seems hard to understand. Creationists argue amongst themselves over the best explanation for this anomaly.

Chapter 34

1. This is a reference to 1 Peter 3:14–16.

Epilogue

1. We should note that the speaker has not accurately described the atheist objection to supernatural explanations. For nearly all atheists, there is no blanket closed-mindedness to the possibility of the supernatural. The point is simply that such explanations, to be credible, must be supported by evidence. The objection to theistic evolution is not that it invokes the supernatural, but that it provides no good reason for doing so.

2. In addition to the other problems with this statement, I would note that this is a further example of two things I discussed in chapter 20. It shows MN being presented as a hard and fast rule that prevents scientists from considering the supernatural, and it also shows MN being used as a shield for protecting religion from science.

3. As is typical at large academic conferences, this was just one session running in parallel with many others. As such, I do not mean to suggest that the various paleontological

societies that organized the convention necessarily approved of, or were even aware of, the substance of these talks. My criticism is directed toward the organizers of the session, who apparently believed this was appropriate fare for a paleontology conference. This is especially ironic given the relentless accusation that atheists were the ones improperly mixing science with religion.

4. I suspect creationists would also argue that modern embryology addresses observable processes going on in the present, while evolution addresses the distant past. For them, that would be another significant difference between embryology and evolution.

BIBLIOGRAPHY

Alexander, Denis. 2008. *Creation or Evolution: Do We Have to Choose?* Oxford: Monarch Books.

Alston, William. 1991. *Perceiving God: The Epistemology of Religious Experience.* Ithaca: Cornell University Press.

Alston, William. 2004. "Religious Experience Justifies Religious Belief." In *Contemporary Debates in Philosophy of Religion*, ed. M. L. Peterson and R. J. Vanarragon. 135–145. Malden: Blackwell Publishing.

Anonymous. 2008. "Evolutionism in the Pulpit" [1911]. In *The Fundamentals*, ed. R. A. Torrey, A. C. Dixon, et al., 4:88–98. Grand Rapids: Baker Books.

Antes, Peter. 2002. "What Do We Experience If We Have Religious Experience?" *Numen* 43.3: 336–342.

Archer, Gleason. 2001. *The New International Encyclopedia of Bible Difficulties.* Grand Rapids: Zondervan.

Artigas, M., T. F. Glick, and R. Martínez. 2006. *Negotiating Darwin: The Vatican Confronts Evolution, 1877–1902.* Baltimore: Johns Hopkins University Press.

Ashton, John F., ed. 2000. *In Six Days: Why Fifty Scientists Choose to Believe in Creation.* Green Forest: Master Books.

Asimov, Isaac. 1983. *The Roving Mind.* Amherst: Prometheus Books.

Asma, Stephen T. 2011. "Risen Apes and Fallen Angles: The New Museology of Human Origins." *Curator* 54.2 (April): 141–163.

Avise, John. 2010. *Inside the Human Genome: A Case for Non-Intelligent Design.* New York: Oxford University Press.

Ayala, Francisco. 2007. *Darwin's Gift to Science and Religion.* Washington D.C.: Joseph Henry Press.

Ayala, Francisco. 2008. "Darwin's Gift to Science and Religion: Commentaries and Responses." *Science and Theology* 6.2: 179–196.

Baggini, Julian. 2003. *Atheism: A Very Short Introduction.* New York: Oxford University Press.

Balmer, Randall. 2006. *Mine Eyes Have Seen the Glory: A Journey into the Evangelical Subculture in America.* 4th edition. New York: Oxford University Press.

Barker, Dan. 2008. *Godless: How an Evangelical Preacher Became One of America's Leading Atheists.* Berkeley: Ulysses Press.

Barlow, Nora, ed. 1958. *The Autobiography of Charles Darwin, 1809–1882.* Posted online at http://darwin-online.org.uk/content/frameset?itemID=F1497&viewtype=text&pageseq=1 (accessed November 6, 2011).

Barton, Bernadette. Forthcoming. *Pray the Gay Away: Religion and Homosexuality in the Bible Belt.* New York: New York University Press.

Beall, Todd S. 2008. "Contemporary Hermeneutical Approaches to Genesis 1–11." In *Coming to Grips with Genesis: Biblical Authority and the Age of the Earth*, ed. T. Mortenson and T. Ury, 131–162. Green Forest: Master Books.

Beatty, John. 1995. "The Evolutionary Contingency Thesis." In *Concepts, Theories, and Rationality in the Biological Sciences*, ed. G. Wolters and J. Lennox, 45–81. Pittsburgh: University of Pittsburgh Press.

Beatty, John. 2006. "Replaying Life's Tape." *The Journal of Philosophy* 103.7 (July): 336–362.

Behe, Michael. 1996. *Darwin's Black Box*. New York: Simon and Schuster.

Behe, Michael. 2008. "Can a Scientific Theory Ameliorate a Theological Difficulty?" *Science and Theology* 6.2: 147–152.

Ben-Menahem, Yemima. 1997. "Historical Contingency." *Ratio* 10.2: 99–107.

Bergman, J., and G. Howe. 1990. *"Vestigial Organs" Are Fully Functional*. St. Joseph: Creation Research Society Books.

Berkman, M., and E. Plutzer. 2010. *Evolution, Creationism and the Battle to Control America's Classrooms*. Cambridge: Cambridge University Press.

Bishop, G., R. Thomas, J. Wood, and M. Gwon. 2010. "Americans' Scientific Knowledge and Beliefs about Human Evolution in the Year of Darwin." *Reports of the National Center for Science Education* 30.1 (May–June): 16–18.

Blackmore, Susan. 2008. "Searching for Meaning in a Pointless World." *Science and Theology* 6.2: 133–136.Bottaro, A., M. Inlay, and N. Matzke. 2006. "Immunology in the Spotlight at the Dover 'Intelligent Design' Trial." *Nature Immunology* 7.5 (May): 433–435.

Borg, Marcus. 2001. *Reading the Bible Again for the First Time: Taking the Bible Seriously but Not Literally*. San Francisco: Harper Collins.

Bottaro, A., M. Inlay, and N. Matzke. 2006. "Immunology in the Spotlight at the Dover 'Intelligent Design' Trial." *Nature Immunology* 7.5 (May): 433–435.

Boudry, M., S. Blancke, and J. Braeckman. 2010. "How Not to Attack Intelligent Design Creationism: Philosophical Misconceptions About Methodological Naturalism." *Foundations of Science* 15.3: 227–244.

Boyd, Steven W. 2008. "The Genre of Genesis 1:1–2:3: What Means This Text?" In *Coming to Grips with Genesis: Biblical Authority and the Age of the Earth*, ed. T. Mortenson and T. Ury, 163–192. Green Forest: Master Books.

Brauer, M., and D. Brumbaugh. 2001. "Biology Remystified: The Scientific Claims of the New Creationists." In *Intelligent Design Creationism and Its Critics*, ed. R. Pennock, 289–334. Cambridge, MA: MIT Press.

Brauer, M., B. Forrest, and S. Gey. 2005. "Is It Science Yet? Intelligent Design Creationism and the Constitution." *Washington University Law Quarterly* 83.1: 1–149.

Broad, C. D. 1998. "The Argument from Religious Experience" [1930]. Reprinted in *Philosophy of Religion: An Anthology*, ed. Louis Pojman. Belmont: Wadsworth.

Brown, William. 2010. *The Seven Pillars of Creation: The Bible, Science, and the Ecology of Wonder*. New York: Oxford University Press.

Carr, Bernard, ed. 2007. *Universe or Multiverse?* Cambridge: Cambridge University Press.

Carroll, Sean. 2010. *From Eternity to Here: The Quest for the Ultimate Theory of Time*. New York: Dutton.

Carter, Matt. 2001. "Lab Scientists Challenging Darwin: 'Intelligent Design' Theory Supports a Thoughtful Creator." *Tri-Valley Herald*, September 26.

Cash, W. J.. 1941. *The Mind of the South*. New York: Alfred A. Knopf

Cheng, C. H. 1998. "Evolution of the Diverse Antifreeze Proteins." *Current Opinion in Genetics and Development* 8.6: 715–720.

Clouser, Roy. 2001. "Is Theism Compatible With Evolution?" In *Intelligent Design Creationism and Its Critics: Philosophical, Theological and Scientific Perspectives*, ed. R. Pennock, 513–536. Cambridge, MA: MIT Press.

Collins, Robin. 2003. "Evolution and Original Sin." In *Perspectives on an Evolving Creation*, ed. K. Miller, 469–501. Grand Rapids: William B. Eerdmans.

Conway Morris, Simon. 2003. *Life's Solution: Inevitable Humans in a Lonely Universe*. Cambridge: Cambridge University Press.

Coyne, Jerry. 1997. "More Crank Science (Commentary on *Darwin's Black Box* by Michael Behe)." *Boston Review* 22.1: 27–28.

Coyne, Jerry. 2009a. "Seeing and Believing." *The New Republic* 240.1 (February 4): 32–41. Posted at http://www.tnr.com/article/books/seeing-and-believing (accessed November 6, 2011).

Coyne, Jerry. 2009b. *Why Evolution Is True*. New York: Viking.

Coyne, J., and H. A. Orr. 2004. *Speciation*. Sunderland: Sinauer Associates.

Craig, William Lane. 1984. "Professor Mackie and the Kalām Cosmological Argument." *Religious Studies* 20.3 (September): 367–375.

Craig, William Lane. 1997. "In Defense of the Kalām Cosmological Argument." *Faith and Philosophy* 14.2 (April): 236–247.

Craig, W. L., and Q. Smith, 1995. *Theism, Atheism and Big Bang Cosmology*. Oxford: Clarendon Press.

Dalrymple, G. Brent. 2004. *Ancient Earth, Ancient Skies: The Age of Earth and Its Cosmic Surroundings*. Stanford: Stanford University Press.

Darwin, Charles. 1859. *On the Origin of Species by Means of Natural Selection*. Available online at http://www.literature.org/authors/darwin-charles/the-origin-of-species/ (accessed November 6, 2011).

Darwin, Charles. 1958. *The Autobiography of Charles Darwin, 1809–1882* [1887], ed. Nora Barlow. London: Collins.

Dawes, Gregory W. 2009. *Theism and Explanation*. New York: Routledge.

Dawkins, Richard. 1983. "Universal Darwinism." In *Molecules to Men*, ed. D. S. Bendall, 403–425. Cambridge: Cambridge University Press.

Dawkins, Richard. 1996. *River Out of Eden*. New York: Basic Books.

Dawkins, Richard. 2006. *The God Delusion*. New York: Houghton Mifflin.

Dawkins, Richard. 2009. *The Greatest Show on Earth: The Evidence for Evolution*. New York: The Free Press.

de Camp, L. Sprague. 1968. *The Great Monkey Trial*. New York: Doubleday.

Dembski, William. 2002. *No Free Lunch: Why Specified Complexity Cannot Be Purchased without Intelligence*. Lanham: Rowman and Littlefield.

Dembski, William. 2004. *The Design Revolution: Answering the Toughest Questions About Intelligent Design*. Downer's Grove: InterVarsity Press.

Dembski, William. 2009. *The End of Christianity: Finding a Good God in an Evil World*. Nashville: B and H Publishing Group.

Dennett, Daniel. 1995. *Darwin's Dangerous Idea: Evolution and the Meanings of Life*. New York: Touchstone.

Dennett, Daniel. 2006. "Atheism and Evolution." In *The Cambridge Companion to Atheism*, ed. M. Martin, 135–148. Cambridge: Cambridge University Press.

DesAutels, Lane. 2009. "Sober and Elgin on Laws of Biology: A Critique." *Biology and Philosophy* 25.2: 249–256.

Desmond, A., and J. Moore. 2009. *Darwin's Sacred Cause: How a Hatred of Slavery Shaped Darwin's Views on Human Evolution*. New York: Houghton Mifflin.

Dixon, Thomas. 2008. *Science and Religion: A Very Short Introduction*. New York: Oxford University Press.

Dobzhansky, Theodosius. 1973. "Nothing in Biology Makes Sense Except in the Light of Evolution." *The American Biology Teacher* 35.3 (March): 125–129.

Domning, D., and M. Hellwig. 2006. *Original Selfishness: Original Sin and Evil in the Light of Evolution*. Hampshire: Ashgate.

Doran, Chris. 2009. "From Atheism to Theodicy to Intelligent Design: Responding to the Work of Francisco J. Ayala." *Theology and Science* 7.4 (November): 337–344.

Drange, Theodore. 1998. *Nonbelief and Evil: Two Arguments for the Nonexistence of God*. Amherst: Prometheus Books.

Draper, K., P. Draper, and J. Pust. 2007. "Probabilistic Arguments for Multiple Universes." *Pacific Philosophical Quarterly* 88: 288–307.

Draper, Paul. 1998. "Evolution and the Problem of Evil." In *Philosophy of Religion: An Anthology*, ed. L. Pojman. Belmont: Wadsworth Publishing.

Dupré, John. 2003. *Darwin's Legacy: What Evolution Means Today*. New York: Oxford University Press.

Ecklund, E. H. 2010. *Science vs. Religion: What Scientists Really Think*. New York: Oxford University Press.

Edis, Taner. 2006. *Science and Nonbelief*. Amherst: Prometheus Books.

Efron, Noah. 2009. "Myth 9: That Christianity Gave Birth to Modern Science." In *Galileo Goes to Jail and Other Myths about Science and Religion*, ed. R. Numbers. Cambridge, MA: Harvard University Press.

Eldredge, Niles. 1982. *The Monkey Business: A Scientist Looks at Creationism*. New York: Washington Square Press.

Elgin, Mehmet. 2006. "There May Be Strict Empirical Laws in Biology, After All." *Biology and Philosophy* 21.1: 119–134.

Eve, R., and F. Harrold. 1991. *The Creationist Movement in Modern America*. Boston: Twayne Publishers.

Fairbanks, Daniel J. 2008. *Relics of Eden: The Powerful Evidence of Evolution in Human DNA*. Amherst: Prometheus Books.

Fales, Evan. 2004. "Do Mystics See God?" In *Contemporary Debates in Philosophy of Religion*, ed. M. L. Peterson and R. J. Vanarragon., 145–158 Malden: Blackwell Publishing.

Fales, Evan. 2009. "Animadversions on *Kitzmiller v. Dover:* Correct Ruling, Flawed Reasoning." Posted at http://www.infidels.org/library/modern/evan_fales/intelligent-design.html (accessed November 6, 2011).

Fitzgerald, Michael. 2008. "Physicist and Priest: An Interview with John Polkinghorne." *Christian Century* 125.2 (January 29): 30–33.

Forrest, Barbara, and Paul R. Gross. 2004. *Creationism's Trojan Horse: The Wedge of Intelligent Design*. New York: Oxford University Press.

Fosdick, H. E. 1932. *As I See Religion*. New York: Harper and Brothers.

Foley, Jim. 2008. "Fossil Hominids: The Evidence for Human Evolution." Posted at http://www.talkorigins.org/faqs/homs/ (last updated June 2010; accessed November 6, 2011).

Frye, Roland. 1983. *Is God a Creationist? The Religious Case against Creation Science*. New York: Scribner.

Futuyma, Douglas. 1995. *Science on Trial: The Case for Evolution*. Sunderland: Sinauer Associates.
Gale, Richard. 1994. "Swinburne's Argument from Religious Experience." In *Reason and the Christian Religion: Essays in Honour of Richard Swinburne*, ed. A. Padgett, 39–65. New York: Oxford University Press.
Gale, Richard. 2007. "The Problem of Evil." In *The Routledge Companion to the Philosophy of Religion*, ed. C. Meister and P. Copan, 397–407. London: Routledge.
Gallup. 2010. "Evolution, Creationism, Intelligent Design." Posted at http://www.gallup.com/poll/21814/evolution-creationism-intelligent-design.aspx (posted in December 2010; accessed November 6, 2011).
Gelinas, Luke. 2009a. "The Problem of Natural Evil I: General Theistic Replies." *Philosophy Compass* 4.3: 533–559.
Gelinas, Luke. 2009b. "The Problem of Natural Evil II: Hybrid Replies." *Philosophy Compass* 4.3: 560–574.
Giberson, Karl. 2008. *Saving Darwin: How to Be a Christian and Believe in Evolution*. New York: HarperOne.
Gilkey, Langdon. 1965. *Maker of Heaven and Earth: A Study of the Christian Doctrine of Creation*. New York: Anchor Books.
Gish, Duane. 1989. "Is It Possible to Be a Christian and an Evolutionist?" *Creation* 11.4: 21–23.
Gishlick, Alan. 2004. "Evolutionary Paths to Irreducible Systems: The Avian Flight Apparatus." In *Why Intelligent Design Fails*, ed. M. Young and T. Edis, 58–71. New Brunswick: Rutgers University Press.
Gishlick, Alan. 2008. "Icons of Evolution? Why Much of What Jonathan Wells Writes about Evolution Is Wrong." Posted at http://ncse.com/creationism/analysis/icons-evolution (last modified October 2008; accessed November 6, 2011).
Gitt, Werner. 2006. *In the Beginning Was Information*. Green Forest: Master Books.
Godfrey-Smith, Peter. 2001. "Information and the Argument from Design." In *Intelligent Design Creationism and Its Critics: Philosophical, Theological and Scientific Perspectives*, ed. R. Pennock. Cambridge, MA: MIT Press.
Gould, Stephen Jay. 1977. *Ontogeny and Phylogeny*. Cambridge, MA: Belknap Press.
Gould, Stephen Jay. 1980. *The Panda's Thumb: More Reflections in Natural History*. New York: W. W. Norton.
Gould, Stephen Jay. 1984. "Human Equality Is a Contingent Fact of History." *Natural History* 93 (November): 26–33.
Gould, Stephen Jay. 1989. *Wonderful Life: The Burgess Shale and the Nature of History*. New York: W. W. Norton.
Gould, Stephen Jay. 1993. "An Earful of Jaw." In *Eight Little Piggies: Reflections in Natural History*, 95–108. New York: W. W. Norton.
Gould, Stephen Jay. 1998. "Showdown on the Burgess Shale: The Reply." *Natural History* 107.10: 48–55.
Gould, Stephen Jay. 1999. "Dorothy, It's Really Oz: A Pro-Creationist Decision in Kansas Is More Than a Blow against Darwin." *Time* 154 (August 23): 59.
Greene, Brian. 2011. *The Hidden Reality: Parallel Universes and the Deep Laws of the Cosmos*. New York: Knopf.
Gregory, Frederick. 1986. "The Impact of Darwinian Evolution on Protestant Theology in the Nineteenth Century." In *God and Nature: Historical Essays on the Encounter*

between Christianity and Science, ed. D. C. Lindberg and R. Numbers. Berkeley: University of California Press.

Griffiths, Paul. 2001. "Genetic Information: A Metaphor in Search of a Theory." *Philosophy of Science* 68.3 (September): 394–412.

Gross, Barry R. 1983."Commentary: Philosophers at the Bar—Some Reasons for Restraint." *Science, Technology and Human Values* 8.4 (fall): 30–38.

Grünbaum, Adolf. 1991. "Creation as a Pseudo-Explanation in Current Physical Cosmology." *Erkenntnis* 35.1/3 (July): 233–254.

Gyatso, Tenzin. 2010. "Many Faiths, One Truth." *The New York Times*, May 24. Posted at http://www.nytimes.com/2010/05/25/opinion/25gyatso.html (accessed November 6, 2011).

Haarsma, Deborah. 2003. "Worshipping the God of Providence." In *Perspectives on an Evolving Creation*, ed. K. Miller. Grand Rapids: William B. Eerdmans.

Hacking, Ian. 1987."The Inverse Gambler's Fallacy: The Argument from Design. The Anthropic Principle Applied to Wheeler Universes." *Mind*, new series, 96.383: 331–340.

Hague, Dyson. 2008. "At-One-Ment by Propitiation." [1910]. In *The Fundamentals*, ed. R. A. Torrey, A. C. Dixon, et al., 3:78–97. Grand Rapids: Baker Books.

Ham, Ken. 1987. *The Lie: Evolution*. Green Forest: Master Books.

Ham, Ken. 2007. *How Could a Loving God . . . Powerful Answers on Suffering and Loss*. Green Forest: Master Books.

Ham, Ken. 2010. *Demolishing Supposed Bible Contradictions: Exploring Forty Alleged Contradictions*. Green Forest: Master Books.

Ham, Ken. 2011. "Does the Gospel Depend on a Young Earth?" *Answers* 6.1 (January–March): 39–40.

Ham, Ken, and A. Charles Ware. 2007. *Darwin's Plantation: Evolution's Racist Roots*. Green Forest: Master Books.

Hanegraaf, Hank. 1998. *The FACE That Demonstrates the FARCE of Evolution*. Nashville: Word Publishing.

Harding, Susan F. 1987. "Convicted by the Holy Spirit: The Rhetoric of Fundamental Baptist Conversion." *American Ethnologist* 14.1 (February): 167–181.

Harlow, Daniel. 2010. "After Adam: Reading Genesis in an Age of Evolutionary Science." *Perspectives on Science and Christian Faith* 62.3 (September): 179–195.

Haught, John F. 2000. *God after Darwin: A Theology of Evolution*. Boulder: Westview Press.

Haught, John F. 2008. *God and the New Atheism: A Critical Response to Dawkins, Harris and Hitchens*. Louisville: Westminster John Knox Press.

Haynes, Stephen R. 2002. *Noah's Curse: The Biblical Justification of Slavery*. New York: Oxford University Press.

Hess, Peter. 2010. "Creation, Design and Evolution: Can Science Discover or Eliminate God?" *University of St. Thomas Journal of Law and Public Policy* 4.1: 102–116.

Hick, John. 1966. *Evil and the God of Love*. London: MacMillan.

Hodge, Charles 1874. *What Is Darwinism?* New York: Scribner, Armstrong.

Holder, R. 2002. "Fine-Tuning, Multiple-Universe and Theism." *Noûs* 36.2 (June): 295–312.

Howard-Snyder, D. 1996. *The Evidential Argument from Evil*. Bloomington: Indiana University Press.

Hull, David. 1991. "The God of the Galapagos." *Nature* 352.8 (August): 485–486.

Hume, David. 1779. *Dialogues Concerning Natural Religion*. Posted online at http://www.anselm.edu/homepage/dbanach/dnr.htm (last accessed November 6, 2011).

Humes, Edward. 2007. *Monkey Girl: Evolution, Education, Religion, and the Battle for America's Soul*. New York: HarperCollins.

Irons, Peter. 2007. *God on Trial: Landmark Cases from America's Religious Battlefields*. New York: Viking Penguin.

Isaak, Mark. 2005. *The Counter-Creationism Handbook*. Berkeley: University of California Press.

Jackelén, Antje. 2008. "What Theology Can Do for Science." *Theology and Science* 6.3: 287–303.

James, William. 1997. *The Varieties of Religious Experience: A Study in Human Nature* [1902]. New York: Simon and Schuster.

Jeffreys, William. 2005. "The Same Old Shell Game." *Reports of the National Center for Science Education*. 25.1–2 (January–April): 47–49.

Johnson, Phillip. 1990. "Evolution as Dogma." *First Things* (October): 15–22.

Johnson, Phillip. 1991. *Darwin on Trial*. Washington D.C.: Regnery Gateway.

Johnson, Phillip. 2000. *The Wedge of Truth: Splitting the Foundations of Naturalism*. Downer's Grove: InterVarsity Press.

Kane, G. Stanley. 1975. "The Failure of Soul-Making Theodicy." *International Journal for Philosophy of Religion* 6.1 (spring): 1–22.

Kardong, Kenneth. 2005. *An Introduction to Biological Evolution*. Boston: McGraw Hill.

Kemp, Kenneth. 2011. "Science, Theology, and Monogenesis." *American Catholic Philosophical Quarterly* 85.2: 217–236.

Kennard, Doug. 2008. "A Nuanced Lakatos Philosophy of Theology and Science." In *Proceedings of the Sixth International Conference on Creationism*, ed. A. Snelling, 31–41. Pittsburgh: Creation Science Fellowship.

Kerby, Carl. 2006. *Remote Control: The Power of Hollywood on Today's Culture*. Green Forest: Master Books.

Kimura, Motoo. 1961. "Natural Selection as the Process of Accumulating Genetic Information in Adaptive Evolution." *Genetics Research* 2.1: 127–140.

Kitcher, Philip 1982. *Abusing Science: The Case against Creationism*. Cambridge, MA: MIT Press.

Kitcher, Philip. 2007. *Living with Darwin: Evolution, Design, and the Future of Faith*. New York: Oxford University Press.

Kitcher, Philip. 2011. "Militant Modern Atheism." *Journal of Applied Philosophy* 28.1: 1–13.

Klein, Richard. 2009. *The Human Career: Human Biological and Cultural Origins*. 3rd edition. Chicago: University of Chicago Press.

Kretzmann, Norman. 1994. "Mystical Perception: St. Teresa, William Alston, and the Broadminded Atheist." In *Reason and the Christian Religion: Essays in Hounour of Richard Swinburne*, ed. A. Padgett, 65–90. New York: Oxford University Press.

Kugel, James L. 1997. *The Bible as It Was*. Cambridge, MA: Harvard Belknap.

Kunzman, Robert. 2010. *Write These Laws on Your Children: Inside the World of Conservative Christian Homeschooling*. Boston: Beacon Press.

Lamoureux, Denis. 2009. *I Love Jesus and I Accept Evolution*. Eugene: Wipf and Stock.

Larson, Edward. 1989. *Trial and Error: The American Controversy over Creation and Evolution*. Updated edition. New York: Oxford University Press.

Larson, Edward. 1997. *Summer for the Gods: The Scopes Trial and America's Continuing Debate over Science and Religion*. New York: Basic Books.

Laudan, Larry. 1982. "Science at the Bar—Causes for Concern." *Science, Technology and Human Values* 7.41: 16–19.

Lawson, A., and W. Worsnop. 1992. "Learning about Evolution and Rejecting a Belief in Special Creation: Effects of Reflective Reasoning Skill, Prior Knowledge, Prior Belief and Religious Commitment." *Journal of Research in Science Education* 39.2: 143–166.

Le Poidevin, Robin. 1996. *Arguing for Atheism: An Introduction to the Philosophy of Religion*. London: Routledge.

Lebo, Lauri. 2008. *The Devil in Dover: An Insider's Story of Dogma v. Darwin in Small-Town America*. New York: The New Press.

Leslie, John. 1988. "No Inverse Gambler's Fallacy in Cosmology." *Mind*, new series, 97.386 (April): 269–272.

Levitt, Norman. 1999. *Prometheus Bedeviled: Science and the Contradictions of Contemporary Culture*. New Brunswick: Rutgers University Press.

Lindberg, David. 2009. "Myth 1: That the Rise of Christianity Was Responsible for the Demise of Ancient Science." In *Galileo Goes to Jail: And Other Myths about Science and Religion*, ed. R. Numbers, 8–18. Cambridge, MA: Harvard University Press.

Livingstone, David. 2008. *Adam's Ancestors: Race, Religion, and the Politics of Human Origins*, Baltimore: The Johns Hopkins University Press

Logsdon Jr., J. M., and W. F. Doolittle. 1997. "Origin of Antifreeze Protein Genes: A Cool Tale in Molecular Evolution." *Proceedings of the National Academy of Sciences* 94.8: 3485–3487.

Long, Eugene Thomas. 1991. "Experience and Natural Theology." *Philosophy of Religion* 31.2/3: 119–132.

Loop, Katherine. 2005. *Beyond Numbers: A Practical Guide to Teaching Math Biblically*. Fairfax: Christian Perspective.

Losos, J. B., T. R. Jackman, A. Larson, K. de Queiroz, and L. Rodriguez-Schettino. 1998. "Contingency and Determinism in Replicated Adaptive Radiations of Island Lizards." *Science* 279.5359: 2115–2118.

Mackie, J. L. 1955. "Evil and Omnipotence." *Mind*, new series, 64.254 (April): 200–212.

Mackie, J. L. 1982. *The Miracle of Theism: Arguments for and against the Existence of God*. New York: Oxford University Press.

Manson, Neil. 2000. "There Is No Adequate Definition of 'Fine-Tuned' for Life." *Inquiry* 43.3: 341–351.

Manson, N., and M. Thrush. 2003. "Fine-Tuning, Multiple Universes, and the 'This Universe' Objection." *Pacific Philosophical Quarterly* 84: 67–83.

Martin, C. B. 1952. "A Religious Way of Knowing." *Mind*, new series, 61.244 (October): 497–512.

Martin, M. 1988. "Reichenbach on Natural Evil." *Religious Studies* 24.1 (March): 91–99.

Martin, M., and R. Monnier, eds. 2003. *The Impossibility of God*. Amherst: Prometheus Books.

Matthews, Mark. 2008. "Evidence for an Earth-Centered Universe." In *Proceedings of the Sixth International Conference on Creationism*, ed. A. Snelling, 231–246. Dallas: Institute for Creation Research.

Matzke, Nick. 2004. "Icons of Obfuscation." Posted at http://www.talkorigins.org/faqs/wells/iconob.html (last modified January 2004; accessed November 6, 2011).

Matzke, Nick. 2009. "But Isn't It Creationism? The Beginnings of 'Intelligent Design' in the Midst of the *Arkansas* and *Louisiana* Litigation." In *But Is It Science: The Philosophical Question in the Creation/Evolution Controversy*, ed. R. T. Pennock and M. Ruse, 377–413. Amherst: Prometheus Books.

Maynard Smith, J., and E. Szathmáry. 1995. *The Major Transitions in Evolution*. New York: Oxford University Press.

McCloskey, H. J. 1960. "God and Evil." *The Philosophical Quarterly* 10.39: 97–114.

McGrath, P. J. 1988. "The Inverse Gambler's Fallacy and Cosmology—A Reply to Hacking." *Mind*, new series, 97.386 (April): 265–268.

McGrew, T., L. McGrew, and E. Vestrup. 2001. "Probabilities and the Fine-Tuning Argument: A Sceptical View." *Mind*, new series, 110.440 (October): 1027–1037.

McIver, Tom. 1994. "The Protocols of Creationism: Racism, Anti-Semitism and White Supremacy in Christian Fundamentalism." *Skeptic* 2.4: 76–87.

McLean, Murdith. 1991. "Residual Natural Evil and Anthropic Reasoning." *Religious Studies* 27.2 (June): 173–188.

McNamara, Patrick, ed. 2006. *Where God and Science Meet: How Brain and Evolutionary Studies Alter Our Understanding of Religion*. 3 vols. Westport: Praeger.

Melendez-Hevia, E., T. G. Waddell, and M. Cascante. 1996. "The Puzzle of the Krebs Citric Acid Cycle: Assembling the Pieces of Chemically Feasible Reactions, and Opportunism in the Design of Metabolic Pathways During Evolution." *Journal of Molecular Evolution* 43.3 (September): 293–303.

Mellert, Robert. 1975. *What Is Process Theology?* New York: Paulist Press.

Miles, Sara Joan. 2001. "Charles Darwin and Asa Gray Discuss Telology and Design." *Perspectives on Science and Christian Faith* 53: 196–201.

Miller, Keith. 2009. "The Misguided Attack on Methodological Naturalism." In *For the Rock Record: Geologists on Intelligent Design*, ed. J. S. Schneiderman and W. D. Allmon. Berkeley: University of California Press.

Miller, Kenneth R. 1999. *Finding Darwin's God: A Scientist's Search for Common Ground Between God and Evolution*. New York: HarperCollins.

Miller, Kenneth R. 2008. *Only a Theory: Evolution and the Battle for America's Soul*. New York: Viking Adult.

Moberly, Walter. 2009. "How Should One Read the Early Chapters of Genesis?" In *Reading Genesis after Darwin*, ed. S. Barton and D. Wilkinson, 5–21. New York: Oxford University Press.

Monton, Bradley. 2009. *Seeking God in Science: An Atheist Defends Intelligent Design*. Peterborough, Ontario: Broadview Press.

Mooney, Chris, and Matthew C. Nisbet. 2005. "Undoing Darwin." *The Columbia Journalism Review* 44.3 (September/October): 30–39.

Moore, James R. 1979. *The Post-Darwinian Controversies: A Study of the Protestant Struggle to Come to Terms with Darwin in Great Britain and America, 1870–1900*. Cambridge: Cambridge University Press.

Moore, John Morrison. 1938. *Theories of Religious Experience*. New York: Round Table Press.

Moore, Randy. 2001. "Racism, Creationism, and the Confederate Flag." *Negro Educational Review* 52.1: 219–228.

Moritz, Joshua. 2008. "Science and Religion: A Fundamental Face-Off, or Is There a *Tertium Quid?*" *Science and Theology* 6.2: 137–145.

Morris, Henry. 1964. "The Bible *Is* a Textbook of Science." Posted at http://www.icr.org/home/resources/resources_tracts_tbiatos (accessed November 6, 2011).

Morris, Henry. 1997. *That Their Words May Be Used against Them: Quotes from Evolutionists Useful for Creationists*. Green Forest: Master Books.

Morris, Henry. 1999. "Design Is Not Enough!" Posted at http://www.icr.org/article/859/ (accessed November 6, 2011).

Morris, Henry. 2002. "Thermodynamics and Biblical Theology." In *Thermodynamics and the Development of Order*, ed. E. Williams. St. Joseph: Creation Research Society Books.

Mortenson, Terry. 2010. "Exposing a Fundamental Compromise." *Answers* 5.3 (July–September): 63–65.

Mortenson, T., and T. Ury, eds. 2008. *Coming to Grip with Genesis: Biblical Authority and the Age of the Earth*. Green Forest: Master Books.

Muller, Hermann. 1918. "Genetic Variability, Twin Hybrids and Constant Hybrids, in a Case of Balanced Lethal Factors." *Genetics* 3.5 (September): 422–499.

Musgrave, Ian. 2004. "Evolution of the Bacterial Flagellum." In *Why Intelligent Design Fails*, ed. M. Young and T. Edis, 72–84. New Brunswick: Rutgers University Press.

Nelkin, Dorothy. 1982. *The Creation Controversy: Science or Scripture in the Schools*. New York: W. W. Norton.

Nelson, Paul. 1996. "The Role of Theology in Current Evolutionary Reasoning." *Biology and Philosophy* 11.4: 493–517.

Numbers, Ronald. 2006. *The Creationists: From Scientific Creationism to Intelligent Design*. Expanded edition. Cambridge, MA: Harvard University Press.

O'Flaherty, Niall. 2010. "The Rhetorical Strategy of William Paley's *Natural Theology* [1802]: Part 1, William Paley's *Natural Theology* in Context." *Studies in History and Philosophy of Science* 41: 19–25.

Oppy, Graham. 1991. "Craig, Mackie, and the Kalām Cosmological Argument." *Religious Studies* 27.2 (June): 189–197.

Oppy, Graham. 2002. "Arguing about the Kalām Cosmological Argument." *Philo* 5.1: 34–61.

Oppy, Graham. 2003. "Creationism on Trial." *Sophia* 42.2 (October): 113–127.

Orr, H. Allen. 1996. "Darwin v. Intelligent Design (Again)." Posted at http://bostonreview.net/BR21.6/orr.html (accessed November 6, 2011).

Orr, H. Allen, and Jerry Coyne. 1992. "The Genetics of Adaptation: A Reassessment." *The American Naturalist* 140: 725–742.

Orr, James. 2008. "Science and Christian Faith." In *The Fundamentals*, ed. R. A. Torrey, A. C. Dixon, et al., 1:334–347. Grand Rapids: Baker Books.

Paley, William. 2006. *Natural Theology, or Evidence of the Existence and Attributes of the Deity, Collected from the Appearance of Nature* [1802]. Ed. M. Eddy and D. Knight. New York: Oxford University Press.

Pallen, M., and N. Matzke. 2006. "From *The Origin of Species* to the Origin of Bacterial Flagella." *Nature Reviews Microbiology* 4.10: 784–790.

Parsons, Keith. 1990. *God and the Burden of Proof: Plantinga, Swinburne, and the Analytic Defense of Theism*. Amherst: Prometheus Books.

Peacocke, Arthur. 2001. "The Cost of New Life." In *The Work of Love: Creation as Kenosis*, ed. J. Polkinghorne, 21–42. Grand Rapids: William. B. Eerdmans.

Pennock, Robert T. 1999. *Tower of Babel: The Evidence against the New Creationism*. Cambridge, MA: MIT Press.

Pennock, Robert T. 2001. *Intelligent Design Creationism and Its Critics*. Cambridge, MA: MIT Press.

Pennock, Robert T. 2007. "God of the Gaps: The Argument from Ignorance and the Limits of Methodologcial Naturalism." In *Scientists Confront Intelligent Design and Creationism*, ed. A. Petto and L. Godfrey. New York: W. W. Norton.

Pennock, Robert T. 2011. "Can't Philosophers Tell the Difference between Science and Religion? Demarcation Revisited." *Synthese* 178.1: 177–206.

Perakh, Mark. 2003. *Unintelligent Design*. New York: Prometheus Books.

Perakh, Mark. 2004. "There Is a Free Lunch after All: William Dembski's Wrong Answers to Irrelevant Questions." In *Why Intelligent Design Fails: A Scientific Critique of the New Creationism*, ed. M. Young and T. Edis, 153–171. New Brunswick: Rutgers University Press.

Perkins Jr., R. K. 1983. "An Atheistic Argument from the Improvability of the Universe." *Noûs* 17.2 (May): 239–250.

Peters, Richard. 2009. "Theodicic Creationism: It's Membership and Motivations." *Geological Society of London, Special Publications* 310: 317–328.

Petto, Andrew, and Laurie Godfrey, eds. 2007. *Scientists Confront Intelligent Design and Creationism*. New York: W. W. Norton.

Pew Forum on Religion and Public Life. 2008, "U.S. Religious Landscape Survey." Posted at http://religions.pewforum.org/reports (last updated June 2008; accessed November 6, 2011).

Pianka, E., and L. Vitt. 2003. *Lizards: Windows to the Evolution of Diversity*. Berkeley: University of California Press.

Pieret, John, ed. 2006. *The Quote Mine Project*. Posted at http://www.talkorigins.org/faqs/quotes/mine/project.html (last updated October. 2006; accessed November 6, 2011).

Pigliucci, Massimo. 2002. *Denying Evolution: Creationism, Scientism, and the Nature of Science*. Sunderland: Sinauer.

Plantinga, Alvin. 2001a. "The Free-Will Defense" [1971]. Reprinted in *God and the Problem of Evil*, ed. W. Rowe, 91–120. Malden: Blackwell.

Plantinga, Alvin. 2001b. "Methodological Naturalism?" [1997]. Reprinted in *Intelligent Design Creationism and Its Critics*, ed. R. Pennock, 339–362. Cambridge, MA: MIT Press.

Plantinga, R. J., T. R. Thompson, and M. D. Lundberg. 2010. *An Introduction to Christian Theology*. Cambridge: Cambridge University Press.

Powell, Russell. 2007. "Is Convergence More Than an Analogy? Homoplasy and Its Implications for Macroevolutionary Predictability." *Biology and Philosophy* 22.4: 565–578.

Prothero, Donald R. 2003. "Inevitable Humans? Or Hidden Agendas?" *Skeptic* 10.3: 54–57.

Prothero, Donald R. 2007. *Evolution: What the Fossil Say and Why It Matters*. New York: Columbia University Press.

Purdom, Georgia. 2005. "The Intelligent Design Movement—How Intelligent Is It?" In *War of the Worldviews: Powerful Answers for an Evolutionized Culture*, 111–120. Green Forest: Master Books.

Quinn, P. L. 1984. "The Philosopher of Science as Expert Witness." In *Science and Reality: Recent Work in the Philosophy of Science*, ed. J. T. Cushing, C. F. Delancy, and G. M. Gutting, 32–53. South Bend: Notre Dame University Press.

Raup, David. 1989. "Testing the Fossil Record for Evolutionary Progress." In *Evolutionary Progress*, ed. M. Nitecki. Chicago: University of Chicago Press.

Reichenbach, Bruce. 1982. *Evil and a Good God*. New York: Fordham University Press.

Reitan, Eric. 2009. *Is God a Delusion? A Reply to Religion's Cultured Despisers*. Malden: Wiley-Blackwell.

Ridley, Mark. 1996. *Evolution*. 2nd edition. Cambridge: Blackwell Science.

Rogerson, John. 2009. "What Difference Did Darwin Make: The Interpretation of Genesis in the Nineteenth Century." In *Reading Genesis after Darwin*, ed. S. Barton and D. Wilkinson, 75–91. New York: Oxford University Press.

Roose, Kevin. 2009. *The Unlikely Disciple: A Sinner's Semester at America's Holiest University*. New York: Grand Central.

Rosenbaum, Ron. 2010. "An Agnostic Manifesto: At Least We Know What We Don't Know." *Slate*, June 28. Posted at http://www.slate.com/id/2258484/ (accessed November 6, 2011).

Rosenhouse, Jason. 2000. "God and Darwin." *Skeptic* 8.2: 84–88.

Rosenhouse, Jason. 2001. "The Design Detectives." *Skeptic* 8.4: 59–64.

Rosenhouse, Jason. 2002a. "Probability, Optimization Theory and Evolution." *Evolution* 56.8: 1721–1722.

Rosenhouse, Jason. 2002b. "Rhetorical Legerdemain in Intelligent Design Literature." In *Darwin Day: Collection One*, ed. A. Chesworth, 327–338. Albuquerque: Tangled Bank Press.

Rosenhouse, Jason, and Glenn Branch. 2006. "Media Coverage of Intelligent Design." *BioScience* 56.3: 247–252.

Rosin, Hanna. 2007. *God's Harvard: A Christian College on a Mission to Save America*. Orlando: Harcourt Books.

Rowe, William, ed. 2001. *God and the Problem of Evil*. Malden: Blackwell.

Runyon, Theodore. 1991. "The Role of Experience in Religion." *Philosophy of Religion* 31.2/3: 187–194.

Ruse, Michael. 2001a. "Methodological Naturalism under Attack." In *Intelligent Design Creationism and its Critics*, ed. R. Pennock, 363–385. Cambridge, MA: MIT Press.

Ruse, Michael. 2001b. *Can a Darwinian Be a Christian?* Cambridge: Cambridge University Press.

Ruse, Michael. 2009. "Why I Think the New Atheists Are a Bloody Disaster." Beliefnet, August 14. Posted at http://blog.beliefnet.com/scienceandthesacred/2009/08/why-i-think-the-new-atheists-are-a-bloody-disaster.html (accessed November 6, 2011).

Ruse, Michael. 2010. *Science and Spirituality: Making Room for Faith in the Age of Science*. Cambridge: Cambridge University Press.

Ruse, Michael. 2011. "New Atheism: A Disaster Comparable to the Tea Party." *Brainstorm*, March 20. Posted at http://chronicle.com/blogs/brainstorm/new-atheism-a-disaster-comparable-to-the-tea-party/33421 (accessed November 6, 2011).

Russell, Bertrand. 1997. *Religion and Science* [1935]. New York: Oxford University Press.

Russell, Bertrand, and F. C. Copleston. 1986. "A Debate on the Existence of God" [1948]. Transcript printed in *Bertrand Russell on God and Religion*, ed. Al Seckel, 123–146. Amherst: Prometheus Books.

Russell, Bruce, and S. Wykstra. 1988. "The Inductive Argument from Evil: A Dialogue." *Philosophical Topics* 16.2 (fall): 133–160.

Russell, Robert John. 2003. "Special Providence and Genetic Mutation: A New Defense of Theistic Evolution." In *Perspectives on an Evolving Creation*, ed. K. Miller. Grand Rapids: William B. Eerdmans.

Russo, Vito. 1981. *The Celluloid Closet: Homosexuality in the Movies*. New York: Harper and Row.

Sarfati, Jonathan. 2002. *Refuting Evolution 2*. Green Forest: Master Books.

Sarkar, Sahotra. 2007. *Doubting Darwin: Creationist Designs on Evolution*. Malden: Blackwell Publishing.

Schwabauer, D., and J. Calvert. 2002. *The Rule: A One-Act Play*. N.p.: The Intelligent Design Network.

Scott, E. C., and G. Branch. 2003. "Antievolutionism: Changes and Continuities." *BioScience* 53.3 (March): 282–285.

Senter, Phil. 2010. "Were Australopithecines Ape-Human Intermediates or Just Apes? A Test of Both Hypotheses Using the 'Lucy' Skeleton." *The American Biology Teacher* 72.2 (February): 70–76.

Shallit, J., and W. Elsberry. 2004. "Playing Games with Probability: Dembski's Complex Specified Information." In *Why Intelligent Design Fails: A Scientific Critique of the New Creationism*, ed. M. Young and T. Edis, 121–138. New Brunswick: Rutgers University Press.

Shanks, Niall. 2004. *God, the Devil and Darwin: A Critique of Intelligent Design Theory*. New York: Oxford University Press.

Sheler, Jeffrey. 2006. *Believers: A Journey into Evangelical America*. New York: Penguin.

Sherbrooke, W. C., A. Scardino, P. de Nys, and L. Schwarzkopf. 2007. "Functional Morphology of Scale Hinges Used to Transport Water: Convergent Drinking Adaptations in Desert Lizards." *Zoomorphology* 126: 89–102.

Shermer, Michael. 2008. "The Chain of Accidents and the Rule of Law: The Role of Contingency and Necessity in the Evolution of Higher Intelligence." *Skeptic* 14.2: 28–36.

Shubin, Neil. 2008. *Your Inner Fish: A Journey into the 3.5 Billion Year History of the Human Body*. New York: Vintage.

Simpson, G. G. 1964. "The Nonprevalence of Humanoids." *Science* 143.3608 (February 21): 769–775.

Singham, Mano. 2009. *God vs. Darwin: The War between Evolution and Creationism in the Classroom*. Lanham: Rowman and Littlefield.

Sinnott-Armstrong, Walter. 2009. *Morality without God*. New York: Oxford University Press.

Smith, Quentin. 2006. "Kalām Cosmological Arguments for Atheism." In *The Cambridge Companion to Atheism*, ed. M. Martin. Cambridge: Cambridge University Press.

Snelling, Andrew. 2009. *Earth's Catastrophic Past: Geology, Creation and the Flood*. 2 vols. Dallas: Institute for Creation Research.

Snoke, David. 2006. *A Biblical Case for an Old Earth*. Grand Rapids: Baker Books.

Sober, Elliott. 1997. "Two Outbreaks of Lawlessness in Recent Philosophy of Biology." *Philosophy of Science* 64.4: S458–S467.

Sober, Elliott. 2003, "It Had to Happen." *The New York Times*, November 20. Posted at http://www.nytimes.com/2003/11/30/books/it-had-to-happen.html (accessed November 6, 2011).

Sober, Elliott. 2004. "The Design Argument." In *The Blackwell Guide to the Philosophy of Religion*, ed. W. Mann, 117–147. Malden: Blackwell.

Sober, Elliott. 2008. *Evidence and Evolution*. Cambridge: Cambridge University Press.

Sober, Elliott. 2010. *Did Darwin Write the Origin Backward? Philosophical Essays on Darwin's Theory*. Amherst: Prometheus Books.

Southgate, Christopher. 2008. *The Groaning of Creation: God, Evolution and the Problem of Evil.* Louisville: Westminster John Knox.

Stenger, Victor. 2011. *The Fallacy of Fine-Tuning: Why the Universe Is Not Designed for Us.* Amherst: Prometheus Books.

Sterelny, Kim. 2005. "Another View of Life." *Studies in History and Philosophy of Biological and Biomedical Sciences* 36.3: 585–593.

Stewart-Williams, Steve. 2010. *Darwin, God and the Meaning of Life: How Evolutionary Theory Undermines Everything You Thought You Knew.* Cambridge: Cambridge University Press.

Swinburne, Richard. 1978. "Natural Evil." *American Philosophical Quarterly* 15.4: 295–301.

Swinburne, Richard. 1996. *Is There a God?* New York: Oxford University Press.

Taylor, J., and J. Raes. 2004. "Duplication and Divergence: The Evolution of New Genes and Old Ideas." *Annual Reviews in Genetics* 38: 615–643.

Tennant, F. R. 1968. *Philosophical Theology* II [1930]. London: Cambridge University Press.

Theobald, Douglas. 2007. "29+ Evidences for Macroevolution: The Scientific Case for Common Descent." Posted at http://www.talkorigins.org/faqs/comdesc/ (accessed February 11, 2011).

Toumey, Christopher. 1994. *God's Own Scientists: Creationists in a Secular World.* New Brunswick: Rutgers University Press.

Trakakis, Nick. 2005. "Is Theism Capable of Accounting for Any Natural Evil at All?" *International Journal for Philosophy of Religion* 57.1: 35–66.

Travisano, M., J. A. Mongold, A. F. Bennett, and R. E. Lenski. 1995. "Experimental Tests of the Roles of Adaptation, Chance, and History in Evolution." *Science* 267.5194 : 87–90.

Turner, Derek. 2011. "Gould's Replay Revisited." *Biology and Philosophy* 26.1: 65–79.

Ward, Keith. 1992. *Defending the Soul.* Oxford: OneWorld Publications.

Ward, Peter. 1991. *On Methuselah's Trail: Living Fossils and the Great Ex*tinctions. New York: W. H. Freeman.

Welch, Gina. 2009. *In the Land of Believers: An Outsider's Extraordinary Journey into the Heart of the Evangelical Church.* New York: Metropolitan Books.

Wells, Jonathan. 1996. "Darwinism: Why I Went for a Second PhD." Posted at http://www.tparents.org/library/unification/talks/wells/darwin.htm (accessed November 6, 2011).

Wells, Jonathan. 2000. *Icons of Evolution, Science or Myth? Why Much of What We Teach about Evolution Is Wrong.* Washington D.C.: Regnery Publishing.

Whitaker, M. A. B. 1988. "On Hacking's Criticism of the Wheeler Anthropic Principle." *Mind*, new series, 97.386 (April): 259–264.

Whitcomb, John C., and Henry M. Morris. 1961. *The Genesis Flood: The Biblical Record and Its Scientific Implications.* Phillipsburg: Presbyterian and Reformed Publishing.

White, Roger. 2000. "Fine-Tuning and Multiple Universes." *Noûs* 34.2: 260–276.

Whitney, Barry L. 1989. *What Are They Saying about God and Evil?* New York: Paulist Press.

Wilkinson, David. 2009. "Reading Genesis 1–3 in the Light of Modern Science." In *Reading Genesis after Darwin*, ed. S. Barton and D. Wilkinson, 127–144. New York: Oxford University Press.

Wise, Kurt. 2002. *Faith, Form and Time: What the Bible Teaches and Science Confirms about Creation and the Age of the Universe.* Nashville: Broadman and Holman.

Wolf, Gary. 2006. "The Church of the Non-Believers," *Wired* 14.11 (available online at http://www.wired.com/wired/archive/14.11/atheism.html; last accessed November 6, 2011).

Wolpe, David. 2008. *Why Faith Matters*. New York: HarperOne.

Wright, George Frederick. 2008. "The Passing of Evolution." In *The Fundamentals*, ed. R. A. Torrey, A. C. Dixon, et al., 4:72–87. Grand Rapids: Baker Books.

Wyse Jackson, Patrick. 2006. *The Chronologer's Quest: The Search for the Age of the Earth*. Cambridge: Cambridge University Press.

Young, Matt, and Taner Edis, eds. 2004. *Why Intelligent Design Fails: A Scientific Critique of the New Creationism*. New Brunswick: Rutgers University Press.

Ziegler, David. 2008. "Predicting Evolution: How Likely Is It That Human-Level Intelligence Will Evolve Again?" *Skeptic* 14.2: 24–27.

Zuckerman, Phil. 2008. *Society without God: What the Least Religious Nations Can Tell Us about Contentment*. New York: New York University Press.

INDEX